S0-CVF-058

ECONOMIC TRANSITION IN HUNAN AND SOUTHERN CHINA

This book is a comparative economic analysis of the five provinces of the Central–South Region of China: Guangdong, Guangxi (autonomous region), Henan, Hunan and Hubei. Of particular interest is a study of the impact of the new policy reforms and readjustment measures on these provincial economies. In tracing the influences of these reforms, special attention is given to the questions of labor utilization and the choice and development of technology.

The book examines the fundamental transition in the Chinese economy from the Mao era to the present Deng regime. It deals with institutional and structural reforms first introduced in 1977–8 and later in 1979. These have brought about several major changes: from a "command economy" to "market socialism", from investment to consumption planning, from the collectives/communes to the production teams, households and even independent workers, and from collective farming to individual peasant farming, private plots and sideline activities.

The choice of Hunan for special treatment is explained by the author's first-hand experience resulting from field visits and personal interviews, and its special significance as the province of Mao and his successor, Hua. The book explores whether the Mao–Hua "leftist" influence still prevails in their province, and if it does, whether it hinders the adoption and diffusion of many of the reforms in Hunan compared with the neighboring provinces *free from this influence*.

Dr A. S. Bhalla is Chief of the Technology and Employment Branch, International Labour Office, Geneva. Formerly he was a Visiting Research Associate at the Economic Growth Center, Yale University. He has been Research Officer at the Institute of Economics and Statistics, Oxford University, and University Tutorial Fellow at the University of Delhi.

He is the editor of *Technology and Employment in Industry* and *Towards Global Action for Appropriate Technology*. He has published articles on development planning, investment allocation, technology choice and employment generation in major economic journals.

Economic Transition in Hunan and Southern China

A. S. Bhalla

St. Martin's Press New York

St. Martin's Press, Inc., 175 Fifth Avenue, New York, NY 10010
Printed in Hong Kong
Published in the United Kingdom by The Macmillan Press Ltd.
First published in the United States of America in 1984

ISBN 0–312–23669–7

To
Chiang Hsieh
and
Joseph Stepanek

who aroused my interest in the Chinese economy

Also by the author

TECHNOLOGY AND EMPLOYMENT IN INDUSTRY (*editor*)
TOWARDS GLOBAL ACTION FOR APPROPRIATE TECHNOLOGY (*editor*)
BLENDING OF NEW AND TRADITIONAL TECHNOLOGIES (*editor with D. James and Y. Stevens*)

Contents

Preface x

Maps xiii

Glossary of Chinese Terms xv

List of Tables and Figures xvi

Abbreviations xix

1 The Provincial Economy **1**
Hunan and the Four Neighbouring Provinces 1
Population and employment 2
Investment allocation 5
Land and agriculture 5
Industry 14
Construction 22
Commerce 24
Concluding Remarks 29

2 The New Economic Policy Frame **31**
Agricultural Pricing 32
Investment Reallocation 34
New measures to promote light industry 35
Heavy industry *v.* light industry 37
Decentralised Economic Management 39
Production Responsibility System 43
Labour Employment System 51

3 The Technology Policy Frame **56**
Technology Choice and Investment Allocation 56
Soviet phase 58
The Great Leap 60
Cultural Revolution 61
Post-Mao period 63

Research, Development and Manpower 65
Two-legs strategy in R and D 66
Linking research and production 67
Shopfloor innovations 68
R and D expenditure 71
Scientific manpower 72
Technology Diffusion 74
Administrative Organisation of Science and Technology 79

4 Technology Imports and Foreign Investment 83
A Historical Perspective 83
Share of Foreign Investment 87
A Preliminary Assessment of Impact 89

5 Agricultural Mechanisation 94
A New Policy of Gradual Mechanisation 94
Implications of New Policy 96
Extent of Mechanisation 98
Mechanisation and Employment 105

6 Rural Industrialisation 111
Concepts and Definitions 111
Some Economic Facts 114
Employment and output growth 114
Inputs, Finance and Organisation 118
Labour supply and training 121
Supply of raw materials and equipment 122
Financing 125
Taxes and profits 126
Administrative organisation 127
Creation of New Enterprises 129
Retrospect and Prospect 130

7 Concluding Remarks 133
Competition Between Urban and Rural Industry 134
Exports *v*. Employment 136

Appendices
- *I* *Statistical Tables* 138
- *II* *Notes on Visits to Industrial Enterprises and Communes* 146
- *III* *Hunan's Textile Industry* 159
- *IV* *Farm Machinery Research Institute (Changsha)* 161

Notes and References 164

Bibliography 178

Name Index 191

Subject Index 194

Preface

This book is based on data collected during my two trips to China, to Beijing and Guangzhou (Canton), in August–September 1980 (when I visited a large fertiliser factory, the South China Institute of Technology in Guangzhou, and the State Science and Technology Commission in Beijing); and to Changsha, Xiantan and Zhu Zhou in Hunan and Guilin in Guangxi in November 1981. These two trips were financed by the ILO, by which I am presently employed. However, the views expressed in this book, which is written in my personal capacity, are solely mine and are not necessarily endorsed by the Organisation.

Much of the information was collected during my second trip when I participated in an International Seminar on the Modernisation of Industry Related to Agriculture held in Changsha. I had an invaluable opportunity to interact closely with important Chinese government officials and scholars from Hunan, its neighbouring provinces and from Beijing, as both the foreign and Chinese participants stayed at the same guest house. I was thus able to indulge in lengthy discussions on several issues besides collecting useful primary data.

The Chinese hosts also organised field trips for the seminar participants to factories, communes and research institutes in Changsha, Shaoyang, Xiantan and Zhu Zhou. I visited a number of small and large factories, a commune, and the Hunan Farm Machinery Research Institute in Changsha. My notes on these visits are presented in Appendices II and IV.

The central focus of the book is on the impact of the new economic policies introduced in 1979 on the provincial economies of Hunan and other provinces of the Central–South Region. This impact is examined particularly in respect of the influence of economic reforms on the employment and technology choice aspects. The economy of the Hunan province is placed in the perspective of its neighbours: namely, Henan, Hubei, Guangxi and Guangdong.

Chapter 1 of the book maps out the economy of Hunan since 1976 in the context of the overall economic situation of the Central–South Region. The population, employment, agricultural and industrial structures of Hunan are compared and contrasted with those of the other provinces in the Region. Chapter 2 outlines the new economic policy framework within which economic changes and policy measures at the provincial level need to be examined. It concentrates on the following specific policy reforms: agricultural pricing, investment reallocation towards light industry, decentralised economic management, the production responsibility system, and the labour employment system. In Chapters 3–6 we attempt to examine the impact of decentralised economic management and investment reallocation on choice of technology, agricultural mechanisation, and employment. Chapter 3 deals with the broad policies of technology transformation in a historical perspective. It is followed up by an analysis of technology imports and foreign investment. Chapter 5 deals with agricultural mechanisation, and Chapter 6 with rural industrialisation and employment. The concluding chapter reviews the issue of competition between urban and rural industry. It also raises some questions of a more general nature, pertaining more particularly to foreign trade and employment.

This book could never have been written or completed without the generous and painstaking help by a large number of Chinese friends in answering questions and providing facts. These persons are too numerous for all to be mentioned here. However, I owe a special debt to three of them – Zeng Dechao of the Beijing Institute of Agricultural Mechanisation, Lui Wenwei of the Hunan Bureau of Commune and Brigade Enterprises, and Zhou Guanxuan of the Hunan Bureau of Mechanical Industry, who continued to offer help even after my departure from China. Two other friends, Chiang Hsieh and Joseph Stepanek, to whom this book is dedicated, aroused my interest in China and its economy long before my trips to that country. Joseph Stepanek actually arranged my participation in the Changsha seminar. William Choa, Chiang Hsieh and Shyam Saran helped me with several translations from Chinese into English.

The writing of this book started mainly in March 1982 in the serene and rural environment of Beas, a village in the north of India. A first preliminary draft of the book was completed in August 1982 in the South of France. It was read in full by Chiang Hsieh, Carl Riskin and Shigeru Ishikawa. All of them gave very penetrating and valuable comments and criticisms. Other China specialists such as Dwight Perkins of Harvard University, and Jon Sigurdson of Lund University, read parts of the manuscript. Zeng Dechao reviewed Chapters 2 and 6. Their comments and suggestions led to a substantial revision of the earlier draft and improved its quality considerably. Needless to say, any remaining errors and deficiencies are my sole responsibility.

Finally, I am grateful to Praveen Bhalla, my wife, and my family for patiently suffering from periodic quarantine due to my 'China fever'. Last but not least, Ita Marguet and Gabrielle Thevenon, who typed parts of the manuscript cheerfully, and Josianne Capt who provided research assistance, also deserve my gratitude.

Commugny, Switzerland A. S. BHALLA

Maps

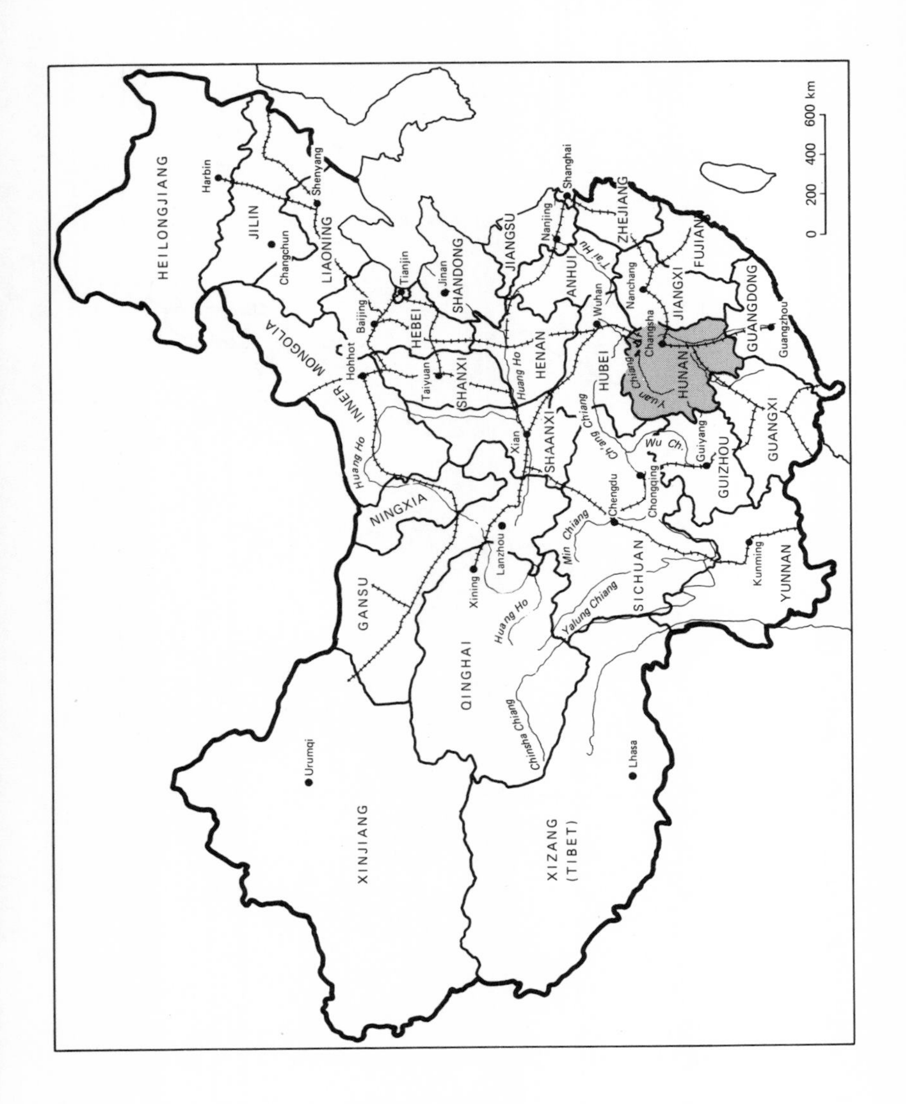

The People's Republic of China *National and provincial boundaries*

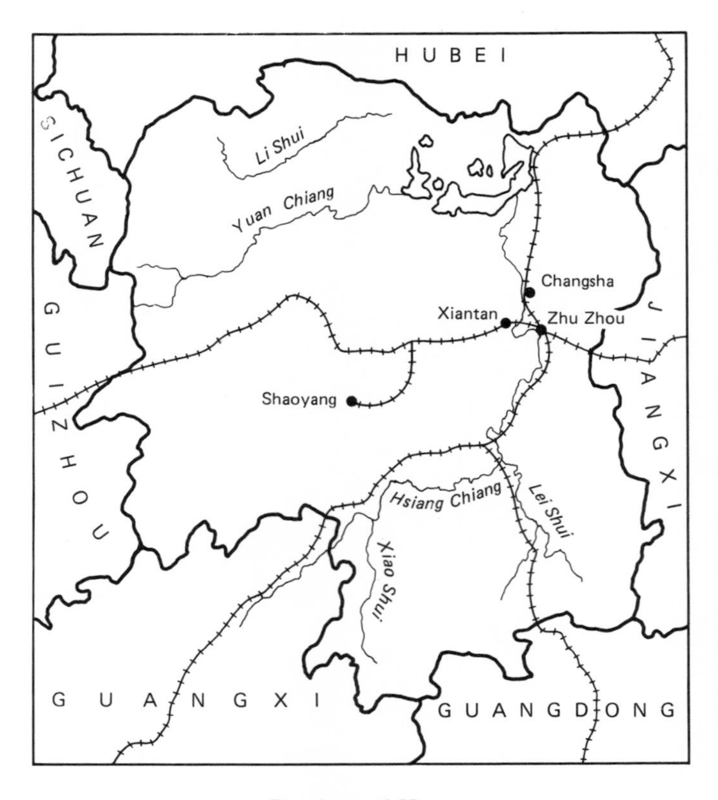

Province of Hunan

Glossary of Chinese Terms

UNITS OF MEASURE

Catty	0.5 kg
Dan	50 kg
Jiao	0.10 yuan
Jin	0.5 kg
Mou or mu	0.0667 hectares or 0.1647 acres
100 jin/mou	0.75 tons/ha
Picul = Dan	50 kg

OTHER TERMS

Hu	Households
Jie Dao	Streets
Lao	Labour
Qu	District
Ren Min Gong She	People's communes
Shedui qiye	Commune and brigade enterprises
Sheng Chan Da Dui	Production brigades
Sheng Chan Dui	Production teams
Xian	County
Zhen	Small town

List of Tables and Figures

TABLES

1.1 Population of the Central–South Provinces (1981) 4
1.2 Sectoral employment: Central–South Provinces (1981) 6
1.3 Sectoral investment in the Central–South Provinces (1981) 7
1.4 Land area and gross agricultural output 8
1.5 Gross value of agricultural output (1981) 9
1.6 Agricultural and industrial output per capita: Central–South Provinces 11
1.7 Hunan's industrial structure 15
1.8 Gross industrial output in the Central–South Region (1981) 18
1.9 Rates of growth of industrial output in the Central–South Provinces (1957–79) 19
1.10 Industrial enterprises by size in the Central–South Region: number, gross output and investment 20
1.11 Light *v.* heavy industry: number, output and investment 21
1.12 Investment, employment and labour productivity in construction: Central–South China (1981)
1.13 Total value of retail sales in the Central–South Region (1981) 24
1.14 System of ownership in retail trade in Hunan and all China (1950–75) 26
1.15 Agricultural price changes in Hunan 28
2.1 Distribution of total gross output of Hunan productions teams (1965–79) 43
2.2 Diffusion of different types of production responsibility system 44
2.3 Adoption of the production responsibility system by form and province 46

2.4 Increase of fixed assets in Wanyu commune (Hunan) under the rural responsibility system 48
3.1 Capital–labour and gross output–labour ratios in Chinese state industrial enterprises 62
3.2 Investment–employment ratios by sector in state-owned units: Central–South Provinces (1981) 64
3.3 Investment in scientific research: Central–South Provinces (1981) 72
3.4 Natural scientific and technical personnel in state-owned units: Central–South Provinces (1981) 73
3.5 Science and technology institutes in Hunan 82
4.1 Share of foreign investment in industrial projects in the Central–South Region 88
4.2 Distribution of investment projects by type of industries (China) 90
5.1 Farm mechanisation by rural institutions 96
5.2 Relative importance of agricultural mechanisation: Central–South Region (1979) 102
5.3 Tractors and power tillers in selected Asian countries 103
5.4 Labour requirements by crops in Hunan (1979) 107
5.5 Input costs by crops in Hunan Province (1979) 109
5.6 Hunan's multiple-cropping index over time (1952–79) 110
6.1 Number, employment and output in commune and brigade enterprises by sector (all China) (1979) 116
6.2 Changes in number, employment and output in commune and brigade enterprises by sector (all China) (1978–9) 117
6.3 Output and employment in commune and brigade enterprises in Hunan Province (1976–80) 119
6.4 Output of commune and brigade enterprises in the Central–South Region (1978–80) 120

APPENDIX TABLES

A.1 Total population of the Central–South Provinces of China 138
A.2 Output of major farm products of the Central–South Provinces 139
A.3 Average unit area yield of major farm crops: Central–South China (1981) 139

A.4 Key agricultural indicators by provinces of the Central–South Region (1979) 140
A.5 Mechanisation: farm machinery stock in the Central–South Provinces (1979–81) 141
A.6 Data on commune and brigade enterprises by provinces of the Central–South Region (1979) 142
A.7 Commune and brigade enterprises: economic data (1976–9) 143
A.8 Gross value of industrial output by provinces of the Central–South Region (1957–79) 144
A.9 Investment in large, medium and small construction projects: Central–South Region (1981) 145
A.10 Basic data on a bicycle factory in Hunan (1981) 147
A.11 Staff of the Institute 161

FIGURES

3.1 Technology variations in China 59
3.2 Administrative structure of science and technology (S and T) in Hunan 81
5.1 Interaction between farm mechanisation and rural commune industry 100
6.1 Rural, 'light' and heavy industry by levels of administration 114

Abbreviations

CCP	Chinese Communist Party
FBIS	*Foreign Broadcast Information Service*
HQ	*Hongqi (Red Flag Biweekly)*
JETRO	Japan External Trade Research Organisation
JJGL	*Jing-ji Guan-li (Economic Management Monthly)*
JJYJ	*Jing-ji Yan-jiu (Economic Research Monthly)*
JMJP	*Jen-min Jih-pao (People's Daily)*
NCNA	*New China News Agency*
OECD	Organisation for Economic Cooperation and Development
RMRB	*Ren-min Ri-bao (People's Daily)*
SCMP	*Survey of China Mainland Press* (Hong Kong)
SCR	State Council Reports
SSB	State Statistical Bureau
SWB	BBC Summary of World Broadcasts
UNCTAD	United Nations Conference on Trade and Development
UNIDO	United Nations Industrial Development Organisation
ZGJJNJ	*Zhongguo Jinji Nianjian* (*Annual Economic Report of China,* 1981)

1 The Provincial Economy

The Central–South Region is one of the administrative and military regions of China which consists of Henan, Hubei, Hunan, Guangxi and Guangdong provinces. These five provinces account for about 270 million of population out of the total Chinese population of 996 million for 1981, or 27 per cent (see Table 1.1). Of this, 27.7 million or 20 per cent is urban population, and 242 million or 28 per cent is rural population. Thus the region is predominantly rural. Of the five provinces, three, namely Guangxi (autonomous region) Henan and Hunan are largely agricultural. While Guangxi and Hunan are mainly rice-growing, Henan is a leading producer of wheat. The remaining two provinces, Guangdong and Hubei, are relatively more industrial. However, in spite of these economic differences between provinces, the region does not form any natural or continguous economic unit. The regional boundaries are arbitrarily drawn on administrative instead of economic lines.

In the pages that follow, we concentrate on the provincial economy of Hunan and examine it in comparison with its member partners in the Central–South Region. There are several reasons for choosing Hunan for special treatment. First, we have had the opportunity to visit the province and obtain first-hand impressions of its economy. Second, it is Mao's and Hua's province and is often dubbed as more 'leftist' than many other provinces. It would therefore be interesting to explore whether this factor plays any significant role in explaining its economic development vis-à-vis that of other provinces. Third, our preliminary fact-finding suggested that the provinces like Hubei and Guangdong were much more intensively studied and reported than Hunan in both Chinese and Western literature.

HUNAN AND THE FOUR NEIGHBOURING PROVINCES

The province of Hunan is situated south of the Yangtse River. It has a territory of 210 000 km^2 of which 62 per cent is accounted for by mountains and plateaus. The province is rich in natural resources, with fertile farmland and a tropical climate. Large tracts of Hunan are covered by ponds, lakes and other ground water suitable for fish breeding. The province is one of the major grain producers in China. Small wonder that it is commonly known as the 'region of fish and rice'.

The province has a mild climate with an annual precipitation of 1300 to 1400 million millimetres. Such climate should, in principle, offer good conditions for a diversified rural economy. But in practice, the mainstay of the province is agriculture and grain production. It is reported that at present only '4.7 per cent of the total agricultural income is provided by forestry, 7 per cent of the province's 3.4 million hectares of cultivated area is sown to industrial crops, and the utilisation of surface water is limited'.[1]

The economy of the province suffered somewhat in 1980 on account of natural disasters like floods. There has been some decline in the production of such agricultural crops as tobacco, jute and ambary hemp, and in the production of oil-yielding crops. This decline may have been partly offset by an increase in the output of commercial and industrial crops like tea, ramie and sugarcane.[2]

Population and Employment

The population of Hunan rose from 33.2 million in 1953 to 35.8 million in 1957, 47.5 million in 1973 and 54 million in 1982 (see Appendix I, Table A.1). This represented a 50.8 per cent increase over the 1957–82 period. This increase is the lowest among all the Central–South Provinces. It is also below the national average increase of 55.9 per cent in the total Chinese population during the same period.

Hunan's population for 1982 is only 5.3 per cent of the total Chinese population of 1008 million. Hunan's share in the total population of China is lower than that of Henan (7.4 per cent) and Guangdong (5.9 per cent), and higher than that of Hubei (4.7 per cent) and Guangxi (3.6 per cent). The share of Hunan in the total population of the Region is 20 per cent compared to 21.8 per cent for Guangdong and 27.3 per cent for Henan, 17.5 per cent for Hubei and 13.4 per cent for Guangxi (see Table 1.1).

The urban population in all the five provinces is quite small, the lowest being accounted for by Guangxi and Hunan which are predominantly agricultural provinces. Even the more industrial provinces like Hubei and Guangdong have only small proportions of the population in the urban areas.

The distribution of the population by sex shows that the males exceed the females in all the five provinces although this increase is less marked in the case of Guangxi.

Data on age-structure of the population for China by province are at present not available, but it would appear that a large proportion of the population consists of youth which makes the problem of youth employment and of the educated unemployed quite serious. This is indicated by frequent references to the problem in the Chinese press. Apart from the natural growth of population, the problem is aggravated by the need to provide jobs for the 'rusticated' youth and demobilised soldiers. For example, in Hunan and other provinces efforts are being made to create jobs for youth through urban collectives and self-employment. Sixty-two counties and municipalities in Hunan have established 193 sugar refineries to arrange jobs for 15 500 urban youth seeking employment.[3] Similarly in Hubei, since the beginning of 1980, special funds of the order of 24.7 million yuan have been allocated for absorbing unemployed youth (including 60 000 rusticated youth), creating jobs for about 224 000 people.[4]

In 1979, some 420 000 people in the cities and towns found employment. In 1980 Hunan province was expected to find jobs for an additional 200 000 people. Towards the end of 1981, the Director of the Hunan Provincial Labour Bureau noted that employment for over 800 000 job-seekers had been created, thus nearly clearing the pre–1979 backlog of unemployed.[5] The share of urban employment is said to have increased in recent years. The bulk of the additional urban jobs has been created through urban collectives and commerce and service activities. For example, in Changsha in 1953 there were 22 000 commerce service points with 46 000 staff and workers. Subsequently, especially during the Cultural Revolution, employment in collective/private commerce was considerably reduced (see below). Since 1979, an attempt has been made to restore old levels in order to ensure employment expansion.[6]

Figures of sectoral employment in state-owned enterprises and urban collectives in the five Central–South provinces and in China as a whole are given in Table 1.2 for 1981. Comparable information for rural commune and brigade enterprises is not available. Yet this partial data shows that in spite of the high priority given to urban collectives, less than 25 per cent of total urban employment in the whole of China is accounted for by them. The state–owned enterprises

TABLE 1.1 *Population of the Central–South Provinces (1981) (million and %)*

Province	*Population in absolute figures and in percentages*														
	Total	*(%)**	*(%)*†	*Male*	*(%)**	*(%)*†	*Female*	*(%)**	*(%)*†	*Urban*	*(%)**	*(%)*†	*Rural*	*(%)**	*(%)*†
Henan	73.9	(7.4)	(27.3)	37.6	(7.3)	(27.2)	36.3	(7.5)	(27.6)	5.8	(4.2)	(20.9)	68.1	(8.0)	(28.1)
Hubei	47.4	(5.0)	(17.5)	24.3	(4.7)	(17.6)	23.0	(4.7)	(17.6)	6.3	(4.5)	(22.8)	41.1	(4.8)	(17.0)
Hunan	53.6	(5.4)	(20.0)	27.8	(5.4)	(20.1)	25.7	(5.3)	(19.6)	5.2	(3.7)	(18.8)	48.4	(5.6)	(20.0)
Guangxi	36.1	(3.6)	(13.4)	18.6	(3.6)	(13.4)	17.5	(3.6)	(13.3)	2.9	(2.0)	(10.5)	33.2	(3.8)	(13.7)
Guangdong	58.8	(5.9)	(21.8)	30.0	(5.8)	(21.7)	28.7	(5.9)	(21.9)	7.5	(5.4)	(27.0)	51.3	(6.0)	(21.2)
Central–South Region	269.8	(27.0)	(100)	138.3	(27.0)	(100)	131.2	(27.0)	(100)	27.7	(20.0)	(100)	242.1	(28.2)	(100)
All China	996.2	(100)		510.8	(51.3)		485.4	(48.7)		138.7	(14.0)		857.5	(86.0)	

* Percentage of the total population of China.
† Percentage of the total population of the Central–South Region.

SOURCE *Statistical Yearbook of China for 1981* (People's Republic of China: SSB, 1982).

remain a more important means of employment generation in China as well as in the five Central–South provinces (see Table 1.2). However, this trend seems to have been reversed in 1982 when, during the first half, 2.4 million jobs are said to have been created in cities and towns out of which 200 000 were in state enterprises and 1 million in urban collectives. Another 100 000 were registered as self–employed; and 1.1 million obtained temporary employment.[7]

Investment Allocation

It is very difficult to make a systematic assessment of sectoral investments in Hunan and the other four provinces of the Central–South Region. Data are not available for collective and private investments which account for a sizeable proportion of investment in the rural sectors of the provincial economies. Information available for investments in state-owned units is presented in Table 1.3 which shows that the bulk of total investments in China as a whole as well as in the five provinces goes to industry with construction and transport accounting for the lowest amounts. Investment figures for agriculture are also fairly low, perhaps partly because collective and private investments are excluded. Fairly sizeable state investments are allocated to public utilities, culture, health, education and social welfare in all the five provinces with the possible exception of Guangxi, which is a relatively poor province.

Land and Agriculture

The total cultivated area in Hunan is 3.44 million hectares which accounts for 3.5 per cent of China's total arable land, and 17.0 per cent of the total arable land in the Central–South Region (See Appendix I, Table A.4). The total irrigated area, 2.44 million hectares, is little over half the total arable area of Hunan. It represents 5.4 per cent of the total irrigated land area in China, and 20.3 per cent of the total irrigated area in the Central–South Region (see Table 1.4 below). At a meeting held in 1981, the Hunan Provincial Water Conservancy Department concluded that only 75 per cent of Hunan's farmland area was well–irrigated and only 50 per cent of the irrigation canals were put to use effectively.[8]

The total volume of grain output for 1979 was 22.18 million tons. Among the Central–South provinces, Hunan accounts for the highest share (6.7 per cent) in China's total grain production. That

TABLE 1.2 *Sectoral employment: Central–South Provinces (1981) (thousand)*

Province	*Total*	*Of which:**		*Sectoral employment*				
		State–owned units	*Urban collectives*	*Agriculture*†	*Industry*	*Construction*	*Transport*‡	*Commerce*§
Henan	4964	4071 (82)	893 (18)	180	2252	378	329	769
Hubei	5316	4216 (79.3)	1100 (20.7)	600	2223	435	380	688
Hunan	4269	3325 (77.9)	944 (22.1)	367	1847	320	328	561
Guangxi	2601	2157 (82.9)	444 (17.1)	247	915	223	226	404
Guangdong	6780	5061 (74.6)	1719 (25.4)	977	2481	396	547	1099
Central–South Region	23930	18830 (78.7)	5100 (21.3)	2371	9717	1752	1810	3521
All China	109397	83722 (76.5)	25675 (23.5)	8614	49017	9245	7212	14962

* Figures in brackets show percentages of the total in each province.
† Includes forestry, animal husbandry and fishing.
‡ Includes posts and telecommunications.
§ Includes catering trade and services.

NOTE Employment includes staff and workers engaged in state-owned units and urban collectives, but excludes those engaged in rural commune and brigade enterprises.

SOURCE *Statistical Yearbook of China for 1981* (People's Republic of China: SSB, 1982).

TABLE 1.3 *Sectoral investment in the Central–South Provinces (1981)* (100 million yuan)*

Province	Sector													
	Agriculture†	*(%)*	*Industry*	*(%)*	*Construction*	*(%)*	*Transport*	*(%)*	*Commerce*	*(%)*	*Other*‡	*(%)*	*Total*	*(%)*
Henan	1.50	(7.9)	11.51	(60.7)	0.32	(1.7)	0.86	(4.5)	1.16	(6.2)	3.62	(19.0)	18.97	(100)
Hubei	1.22	(5.8)	12.54	(59.6)	0.45	(2.1)	0.74	(3.5)	1.32	(6.3)	4.79	(22.7)	21.06	(100)
Hunan	0.81	(6.5)	5.02	(40.3)	0.29	(2.3)	0.66	(5.3)	1.57	(12.6)	4.12	(33.0)	12.47	(100)
Guangxi	0.97	(14.2)	2.97	(43.5)	0.11	(1.6)	0.35	(5.1)	0.51	(7.5)	1.92	(28.1)	6.83	(100)
Guangdong	5.53	(17.4)	9.91	(31.2)	0.28	(0.9)	3.81	(12.0)	3.62	(11.4)	8.64	(27.1)	31.79	(100)
Central–South Region	10.03	(11.0)	41.95	(46.0)	1.45	(1.6)	6.42	(7.0)	8.18	(9.0)	23.19	(25.4)	91.12	(100)
All China	31.71	(7.5)	215.26	(50.3)	9.21	(2.2)	40.47	(9.4)	28.01	(6.5)	103.23	(24.1)	427.89	(100)

* The data relate only to state-owned units. Collective and private investments are excluded.

† Includes investment in geological prospecting.

‡ This category includes civil public utilities, scientific research, culture, health, education and social welfare, financing and insurance, and government agencies and public organisations.

SOURCE *Statistical Yearbook of China for 1981* (People's Republic of China: SSB, 1982).

TABLE 1.4 *Land area and gross agricultural output (percentages)*

Province	*Arable land*	*Irrigated area*	*Volume of grain output*
Share of each province in the total for Central–South Region			
Henan	35.4	30.2	23.4
Hubei	18.6	19.5	20.3
Hunan	17.0	20.3	24.3
Guangxi	13.0	12.1	12.9
Guangdong	15.9	17.9	19.1
Central–South Region	100.0	100.0	100.0
Share of the provinces of the Central–South Region in the total for China			
Henan	7.2	8.1	6.4
Hubei	3.8	5.2	5.6
Hunan	3.5	5.4	6.7
Guangxi	2.6	3.2	3.5
Guangdong	3.2	4.8	5.2
Gentral–South Region	20.3	26.8	27.4
Total China	100.0	100.0	100.0

SOURCE Percentages computed from information supplied by the State Agricultural Commission. For absolute figures, see Appendix I, Table A.4.

Hunan is one of the predominantly agricultural provinces is also indicated by its high share of 24.3 per cent in the total volume of grain output in the Region (see Table 1.4).

By November 1981, Hunan province had crash-harvested late rice on 44 per cent of its 30 million *mou* area.[9] However, the months of October and November 1981 were particularly wet, drizzly and windy. In Changsha, we were told that these bad weather conditions were likely to cause grave damage to the late rice which was ripe and ready to be harvested. (We visited Hunan in the month of November 1981 when most of the paddy fields were wet.) It is estimated that the early rice output in 1981 reached 19 350 million *catties*,[10] that is, an additional 600 million or 3.5 per cent more than the figure for 1980. The average yield (from an area of 29 million *mou*) of early rice was some 660 *catties* per *mou*, that is, 20 *catties* per *mou* more than the figure for 1980.[11]

The total rape–seed output in 1981 exceeded 5 million *piculs*.[12] In the Lianyuan, Shaoyang and Chenzhou Prefectures, the rape-seed output was reported to be three times the figure for 1980. All the 16 prefectures and municipalities of Hunan have raised their rape-seed output: bumper crop harvests are reported in the 105 counties and municipalities. In 39 counties the yield exceeded 100 *catties* per *mou*.[13]

Output Value

As shown in Table 1.5 below, the gross agricultural output of Hunan was the second highest in the Central–South Region after Henan. It accounts for 5.6 per cent of the total agricultural output of China as a whole.

According to the Director of the Hunan Provincial Agricultural Department,

> a comparison of farm production in 1979 with that in the year when the province was liberated shows that the province's total agricultural output value increased 4.9 times, or at an average of 4.5 per cent a year over the past 30 years; total grain output increased 2.5 times, or at an average rate of 4.2 per cent a year; cotton output

TABLE 1.5 *Gross value of agricultural output (1981)*

Province	*Gross value of agricultural output (100 million yuan)**	*Total population (million)*	*Rural population (million)*	*Gross value of agricultural output per capita (yuan)*	
Henan	163.5	73.9	68.1	221.2†	240.0‡
Hubei	114.0	47.4	41.0	240.5†	278.0‡
Hunan	131.4	53.6	48.4	245.1†	271.5‡
Guangxi	71.9	36.1	33.1	199.1†	217.2‡
Guangdong	120.3	58.8	51.3	204.6†	234.5‡
All China	2311.9	996.2	857.5	232.0†	269.6‡

* Includes crop production, fisheries, forestry, animal husbandry and brigade rural enterprises and other sideline activities.
† Divided by total population.
‡ Divided by rural population.

SOURCE SSB, People's Republic of China.

increased 12.4 times, or at an average rate of 9 per cent a year.[14]

For China as a whole, Ishikawa has estimated the average annual rate of growth of total agricultural output at 2.25 per cent for 1957–70 and 3.46 for 1970–7, which is below the rate achieved by Hunan.[15]

The gross value of agricultural output per capita for Hunan is above the national per capita whether one takes account of total population or rural population only. The ranking of different provinces of the Central–South Region by gross value of agricultural output per capita is affected somewhat by the differences in rural and total population (see Table 1.5). However, since the liberation, the average agricultural labour productivity in China has remained quite low in spite of growth in total farm production. It is estimated that the growth of grain output per capita between 1957 and 1979 has been very slow, and in some cases even negative.[16] For China as a whole, the grain output per capita increased from 290 kg in 1957 to 320 kg in 1979, thus accounting for only a 10.3 per cent increase over the whole period. In the Central–South Region, Hubei is the only province showing an increase from 356 kg per capita in 1957 to 400 kg per capita in 1979 (12 per cent). In the cases of Henan and Guangdong, the per capita grain output actually fell during this period. (Figures are not available for Guangxi and Hunan.)

The low agricultural labour productivity seems to explain the low amount of marketed agricultural surplus. The ratio of marketed to total foodgrains output has been estimated by Ishikawa as follows: 25.1 per cent gross and 11.3 per cent net for 1956–7, 12–13 per cent net for 1977 and 1978, and 20 per cent gross and 15 per cent net for 1979 or 1980.[17] This sluggish growth of foodgrains surplus may be explained partly by the inadequate output-raising properties of technological change during this period. According to Ishikawa, although the non-labour inputs (both biological and mechanical) were considerable in absolute terms, they seem to have been 'insufficient for raising labour productivity'. Since marketed food surpluses are important for industrialisation and urbanisation, the Chinese authorities seem quite concerned about raising them through appropriate agricultural development strategies. For raising marketable food surpluses, the Chinese authorities have identified priority provinces as follows: Heilungjiang and Jilin in the North for wheat, soybeans and sugar, and Hunan and Jiangsu in the South for rice, cotton, and oil-yielding crops.[18]

Table 1.6 compares the per capita agricultural output of the Central–South Provinces with their per capita industrial output. The per capita agricultural output is much lower than the per capita industrial output in all the five provinces. The per capita agricultural output of the provinces of Hubei and Hunan exceeds the national per capita agricultural output. But this is not true of the per capita industrial output of these provinces, which is well below the national average. This implies a lower average industrial productivity in the Central–South Provinces.

Rural Diversification

The rural economy of Hunan has been concentrated mainly on the production of foodgrains. Recently, the province has introduced a nation-wide drive to diversify the rural economy. This drive involves the cultivation of vegetables and other non-food crops, and the promotion of forestry and related activities. Generation of electricity in the province should in principle facilitate the development of rural non–farm activities. During the first half of 1981, Hunan generated 5800 million kilowatts of electricity, fulfilling 51.9 per cent of its 1981 quota and achieving an increase of 2.7 per cent over 1980.[19]

The area covered under forestry is being increased. For example,

TABLE 1.6 *Agricultural and industrial output per capita: Central–South Provinces (in 1980 fixed prices)*

Province	*Per capita agricultural output* (yuan)*	*Percentage of national per capita (%)*	*Per capita industrial output† (000 yuan)*	*Percentage of national per capita (%)*
Henan	240	88.8	352	94.3
Hubei	278	103.0	327	87.6
Hunan	271	100.3	338	89.9
Guangxi	217	80.3	283	75.8
Guangdong	234	86.6	333	89.3
National per capita	270	100.0	373	100.0

* Divided by rural population.
† Divided by urban population.

SOURCE SSB, People's Republic of China.

the Zhu Zhou county (which we visited in November 1981) has raised its forested area to nearly 1.15 million *mou* or 43.2 per cent of its total land area. The county has also launched a movement to plant more trees. During the Spring Festival in 1981 an afforestation drive was organised to mobilise masses to create more than 35 000 *mou* of forests by the Spring of 1982.[20]

The jurisdiction between the state forests, collectively owned forests, private plots on hillsides, and scenic forests has been drawn up in such a way that the unit of administration financing the planting of trees reserves the right of ownership. The commune members are allowed to use the trees planted by themselves around their houses. This new policy has reduced deforestation and wanton felling of trees.

In implementing the agricultural responsibility system (see Chapter 2) in the Zhu Zhou county, some communes (for example, Huanglong commune) have also introduced the forestry responsibility system under which contracts have been signed for specialised jobs and remuneration linked to output. As a result, higher yields are reported in all the 27 commune and brigade-operated yards.

The production of vegetables, which are often in short supply, is being given much greater emphasis. A number of special measures have been introduced in order to overcome shortage and the high price of vegetables. First, the marketing of vegetables is now linked to the supply of foodgrains. Formerly the vegetable growers received the same grain ration irrespective of the amount of vegetables sold by them to the state. In order to offer incentives to the vegetable-growing farmers, the grain ration is now calculated on the basis of the amount of vegetables sold (the greater the amount of vegetables sold, the greater the grain ration for the growers). Furthermore the farmers are offered a bonus in the form of grain on top of their basic grain ration if they overfulfil their vegetable production quotas.[21] Second, responsibility and contract systems are being introduced for vegetable growing. Third, according to the provincial plan for 1982 the funds for scientific research in vegetable growing and for water-control construction in vegetable areas will be 200 000 yuan each more than the 1981 allocations. Fourth, the provincial authorities have promised to improve vegetable market management in order to control speculation and profiteering.[22]

Diversification of the Hunan rural economy also covers development of industrial crops and animal husbandry in the form of collective as well as sideline activities by the individual households. The production brigades and individual households are encouraged to spend more time on such non-farm activities as rural processing industries and the diffusion of proven innovations among farmers through local science and technology systems. These two particular issues are treated in Chapters 3 and 6.

Although grain production has risen steadily in Hunan for the past several years, the diversification of the rural economy through forestry, industrial crops and animal husbandry remains rather slow. These non-farm activities represent a small proportion of the total agricultural output of Hunan. However, this slowness of diversification may be due partly to lack of transportation and marketing infrastructure.

Rural Incomes

Some diversification of the rural economy of Hunan has contributed to an increase in the average per capita income of its 40 million peasants. In 1980, the absolute level of per capita income was 93.8 yuan compared to 92.3 yuan for 1979. In a number of communes, the incomes were well above this average, and ranged from 200 to 400 yuan per capita.[23] The average rural income per capita in Hunan is higher than that in Henan (63.4 yuan), Guangxi (74.7 yuan) and Guangdong (88.4 yuan) but lower than that in Hubei (106.2 yuan). It is also above the national average of 83.4 yuan.

A more recent survey of the incomes of 600 households in 20 counties shows that per capita income in the first half of 1981 was 60.67 yuan in cash; that is, 82 per cent more than the cash income in the corresponding period for 1980. It is felt that this increase may have been caused by the introduction of such new rural economic policies as the household responsibility system. The per capita incomes of the households of commune members from sideline occupations in the first half of 1981 was 41.6 yuan, which accounted for 68.5 per cent of their total income. It was 95.6 per cent more than the figure for 1980 for the corresponding period.[24]

Rural Savings

Although adequate data are not available, it would appear that the potential for saving has increased with an increase in rural incomes in the province. This seems to be particularly true of rural savings for investment in household activities which have to be financed almost entirely out of local resources.

Rural savings in China as a whole are reported to have increased by 29 per cent in 1981. In absolute terms, China's rural savings amounted to 25.8 yuan per head, a 7.6 yuan increase over 1980. In the provinces of Guangdong, Liaoning, and suburban areas of Beijing, Shanghai and Tianjin, rural savings per head were as high as 40 yuan or more.[25]

By the end of 1981, the balance of savings deposits at the Agricultural Bank and credit cooperatives reached 21.17 million yuan, exceeding the figure for 1980 by 6246 million yuan, that is, an increase of 41.9 per cent. The personnel savings deposits absorbed by the Chinese credit cooperatives amounted to 16 947 million yuan.[26] In the province of Hunan, at the end of June 1981, rural banking facilities had been extended to communes with agricultural loans totalling 990 million yuan, representing an 18.7 per cent increase over 1980 for the same period.[27] For China as a whole, according to the Agricultural Bank of China, the agricultural loans extended between January and August 1981 amounted to 21 800 million yuan, representing a 23 per cent increase over 1980.

Industry

The total value of Hunan's industrial output in 1980 was 60 times that in 1949. In 1980, industry accounted for 64.4 per cent of the total value of industrial and agricultural production as against 16.7 per cent in 1949.

At the end of 1980, Hunan had 20 440 industrial and transport enterprises employing about 1.74 million people including administrative staff and unskilled workers. The province had another 103 000 commune and brigade enterprises of which 45 000 were industrial enterprises. This latter group of enterprises generated a total income of 3.1 billion yuan (2000 million yuan from the industrial enterprises) and a total employment of 1.5 million, of which 910 000 were engaged in rural industrial enterprises (see Chapter 6).

Industrial Structure

The industrial composition of the Hunan economy is biased in favour of light industry covering the processing of food crops and agricultural raw materials. Table 1.7 gives some indication of the nature and structure of industry in the province.

In Hunan, the handicrafts seem to be largely of the artistic type intended for tourist and export markets (for example, products like

TABLE 1.7 *Hunan's industrial structure*

Handicrafts (utilitarian and artistic)	*Light industries (consumer goods)*		*Heavy industries (capital and intermediate goods)*
	Food	*Non-food*	
Embroidery	Tea-processing	Rice–bran oil extraction	Steel
			Agricultural machinery
Firecrackers	Rice-milling	Silk-reeling	Electronics
Bamboo-carving	Flour-milling	Paper	
Bamboo sheets		Textiles	
Products made of feathers		Pottery and ceramics	
		Leather tanning	
		Electric fans	
		Bicycles	

embroidery, and articles of feathers and carved bamboo). In view of the relatively low purchasing power, there seems to be a stable local market for such utilitarian handicrafts as pottery and ceramics, bamboo sheets and embroidered cloth.

Although the present level of purchasing power is low, it has already risen and is expected to rise still further. For example, an analysis of expenditure on 27 major commercial products showed an increase of 27.5 per cent in 1979 over that in 1978. Of these products, basic commodities for daily requirements showed an increase of 57.8 per cent: clothes, an increase of 49.6 per cent: and food, an increase of 13 per cent. In the capital city of Changsha, purchasing power has increased at an average rate of 13 per cent since 1976: it is expected to grow by 62.4 per cent during the period 1980–5, representing a rate of increase of 10 per cent per annum.[28]

Part of the increase in purchasing power is likely to generate additional demand for handicraft products of the utilitarian type, to the extent that a shift from food to non-food expenditure is likely to occur. However, there are indications that consumers are beginning to prefer urban goods and consumer durables like radios, sewing machines, bicycles and televisions.[29] This suggests that the additional purchasing power may spill over more to raise demand for the products of light industries than for the utilitarian handicrafts.

The province of Hunan is well endowed with raw materials required for the development of light industries. It is rich in pottery, clay, non-ferrous metals, forests, bamboos, pig skin, feathers, and so on. The light industries are at present located mainly in the capital city of Changsha, although it is planned to develop these industries also in other provincial towns.

The main light industries of Hunan are: tea-processing, rice-milling, textiles, ceramics, leather-tanning and bicycles. There are 82 ceramics plants in Hunan including 14 major factories specialising in export production. Liling porcelain, well-known in the country, is produced in five large and medium-sized factories employing 20 000 workers producing mainly for export, and one factory engaged in domestic production. Liling county alone supplies 10 per cent of China's total export production, or about 90 million pieces of underglaze daily use porcelain. In 1981, export value of ceramics from Hunan totalled 24 million US dollars with a total production amounting to 130 million pieces. Since 1980, Chinese porcelain ware has started to capture the US market (thanks to the most-favoured nation treatment) at the expense of Japanese porcelain which has become relatively costly in recent years. It is reported that the 'Chinese share of the US market for low-value dinner-ware has grown from approximately 1.5 per cent in 1977 and 1978 to more than 50 per cent in 1981, and has reached 74.5 per cent in the first quarter of 1982'.[30]

In November 1981, we visited some of the light industry enterprises in Hunan (for example, bicycles and porcelain). The descriptions of these factory visits are given in Appendix II.

The heavy industry of Hunan includes agricultural machinery, industrial pumps, electronics, wires and cables, pistons and piston rings, steel, and so on. Under the new economic policy, the heavy industry is urged to serve agriculture and light industry. The Industry and Transportation Conference of Hunan, held in April 1981, concluded that

> the metallurgical, machine-building, defence, chemical and other industries should adapt themselves to serving agriculture and the textiles and light industries, not only to carry out technical innovations and to provide up-to-date standard raw and semi-finished materials, but to organise joint efforts and cooperation to produce consumer goods demanded by the market.

At the Ninth Session of the Fifth Hunan Provincial People's Congress Standing Committee in May 1981, Li Quangshang, the Chairman of the Provincial Economic Committee, stated that 'it is necessary to fully tap the potential of existing enterprises and to have heavy and light industries support each other so as to speed up the expansion of consumer goods production.'

De-emphasis of heavy industry, under the new economic policies introduced in 1979 (see Chapter 2), has led to large excess capacities in many Hunan factories including those which we visited in November 1981 (see Appendix II for a discussion of these visits). There are also a number of other problems faced by heavy industry. First, the production of high precision machine tools is limited. Second, the product design capability and the control system are poor, which leads to an inefficient use of energy and other inputs. Third, the quality of domestic machinery is poor due to the use of inferior-quality metals and other materials. Finally, much of the machinery in the factories is unguarded; the factories suffer from environmental pollution and poor safety and hygiene conditions.

Our field observations suggest that the light consumer goods industries (such as bicycles and porcelain manufacturing which we visited) also suffer from poor-quality products and the inefficient use of technology

Industrial Output

According to official estimates for 1979, Hunan's gross value of industrial output was 15 617 million yuan (at 1979 prices) compared with 459 245 million yuan for China as a whole and 81 860 million yuan for the Central–South Region. Thus Hunan's share of China's total industrial output is only 3.4 per cent compared to 3.7 per cent for Henan, 4.0 per cent for Hubei, 1.6 per cent for Guangxi, and 4.8 per cent for Guangdong. Within the Central–South Region, Hunan accounts for 19 per cent of the gross regional industrial output which is the lowest share after Guangxi (see Table 1.8 below). This implies that Hunan is one of the relatively less industrialised provinces of China as well as of the Central–South Region.

Some time series data on gross industrial output by provinces is available (see Appendix I, Table A.8) which enabled us to analyse industrial output growth in Central–South China over time. Table 1.9 below gives the annual rates of industrial growth for the provinces of this region for different time-periods since 1957.

During the period before the Cultural Revolution, the annual rate of growth of industrial output was the highest in Henan followed by Guangdong and the second lowest in Hunan after Hubei. Hunan's

TABLE 1.8 *Gross industrial output in the Central–South Region (1981)*

Province	*Gross industrial output (100 million yuan)*			*Percentage share in total China*			*Percentage share in Region*		
	Total	*Light industry*	*Heavy industry*	*Total*	*Light industry*	*Heavy industry*	*Total*	*Light industry*	*Heavy industry*
Henan	20 400	11 200	9 200	3.9	4.2	3.6	22.3	20.0	21.7
Hubei	20 600	12 800	11 800	4.0	4.8	4.7	22.4	23.9	27.8
Hunan	17 600	8 200	9 400	3.4	3.0	3.7	19.2	15.3	22.8
Guangxi	8 200	5 200	3 000	1.6	1.9	1.2	8.9	9.7	7.0
Guangdong	25 000	16 200	8 800	4.8	6.0	3.5	27.2	30.2	20.7
Central–South Region	91 800	53 600	42 200	17.7	20.1	16.7	100.0	100.0	100.0
All China	517 800	266 300	251 500						

SOURCE *Statistical Yearbook of China for 1981* (People's Republic of China: SSB, 1982).

TABLE 1.9 *Rates of growth of industrial output in the Central–South Provinces (1957–79) (%)*

Province	Annual growth				
	(1957–65)	*(1965–70)*	*(1970–5)*	*(1975–9)*	*(1970–9)*
Henan	11.5	13.9	10.8	13.4	11.9
Hubei	6.4	12.9	5.3*	18.8†	12.6
Hunan	7.6	15.5	6.9	21.3	13.1
Guangxi	7.8	13.2	15.2	21.9	18.1
Guangdong	10.3	10.0	10.1	4.9	7.8
All China	8.9	11.7	9.1	9.3	9.2

* 1970–4
† 1974–9

SOURCE Appendix I, Table A.8 on the basis of which these rates have been derived.

growth of industrial output increased significantly during the 1965–70 period when it was indeed the most rapid of that of all the Southern provinces. Industrial growth declined sharply during 1975–9 in Hunan and Hubei although it remained fairly stable in Guangdong and, to a lesser extent, in Henan and Guangxi. If we take the 1970s as a whole, the annual rate of growth of industrial output in Hunan becomes the second highest after Guangxi whereas that in Guangdong becomes the lowest. Although we are uncertain about the reliability and comparability of data, it is plausible that Guangdong, being a more industrial province than the others, had a much higher level of initial industrial output (see Appendix I, Table A.8), and hence its industrial output grew more slowly than that of the other provinces.

The small-scale enterprises in all five provinces accounted for the bulk of the total industrial output, followed by large enterprises. In Henan, Hubei, Hunan and Guangdong, large enterprises accounted for greater industrial output than the medium ones, although in all these cases they were fewer in number. This implies higher productivity of the larger enterprises in these provinces (see Table 1.10).

In 1981, the gross output of light industry in Henan, Hubei, Guangxi and Guangdong was higher than that of heavy industry. In the case of Hunan, however, heavy industry accounted for greater output with a fewer number of enterprises (see Table 1.11).

TABLE 1.10 *Industrial enterprises by size in the Central–South Region: number, gross output and investment*

Province	*Large*			*Medium*		*Small*		
	Number	*Gross output (100 million yuan)*	*Investment (large and medium) (100 million yuan)*	*Number*	*Gross output (100 million yuan)*	*Number*	*Gross output (100 million yuan)*	*Investment (100 million yuan)*
Henan	59	37.7	9.38	98	36.7	15 736	129.0	9.60
Hubei	59	78.9	8.29	168	42.7	16 805	125.0	12.77
Hunan	57	39.0	2.22	118	27.0	19 842	110.2	10.25
Guangxi	21	4.8	1.82	113	19.7	10 393	57.0	5.01
Guangdong	44	36.4	5.52	175	31.7	23 028	182.0	26.27
Central–South Region	240	196.8	27.23	672	157.8	85 804	603.2	63.90
All China	1 476	1 316.6	169.18	3 544	915.8	376 523	2 945.3	252.55

SOURCE *Statistical Yearbook of China for 1981* (People's Republic of China: SSB, 1982).

TABLE 1.11 *Light v. heavy industry: number, output and investment*

Province	*Number (000)*		*Gross output (100 million yuan)*		*Investment (100 million yuan)*	
	Light	*Heavy*	*Light*	*Heavy*	*Light*	*Heavy*
Henan	9.4	6.5	111.7	91.8	4.10	7.41
Hubei	11.2	5.8	128.0	118.6	2.49	10.05
Hunan	11.4	8.6	81.6	94.6	1.07	3.95
Guangxi	7.1	3.4	52.2	29.4	0.68	2.29
Guangdong	16.3	6.9	162.0	88.4	3.33	6.58
Central–South Region	55.4	31.2	535.5	422.8	11.67	30.28
All China	235.9	145.6	2662.9	2514.8	42.6	172.6

SOURCE SSB, People's Republic of China.

Industrial Investment

In general, prior to 1977 the Chinese industrial structure was quite unbalanced with undue emphasis on heavy industry at the expense of light industry. From 1949 to 1979 total investment in light industry was much smaller than that in heavy industry.

It is reported that 'between 1966 and 1978, of the total amount of investment in capital construction, heavy industry took up more than 55 per cent, agriculture a little more than 10 per cent and light industry even less, only about 5 per cent'.[31] Neglect of light industry is also explained by the fact that in 1976, only 2.1 per cent of China's machinery output was built 'by heavy industry (which included mining and electric power) for light industry' and in 1978, only 11.7 per cent of the steel products made in China were used for light industrial goods. In 1981, investment in light industry in China as a whole remained about one-quarter of that in heavy industry. Similarly, in each of the Central–South Provinces it was relatively quite limited. However, the gross output of light industry was higher in all the provinces except Hunan.

It is likely that the slowing down of the Chinese policy of Four Modernisations (that is, long-term agricultural, industrial, military, and scientific and technological development) in the Spring of 1979 has also slowed down industrial growth in Hunan and the other four provinces. Realising that the initial targets for 1985 were too ambitious, in 1979 the Chinese leadership reduced the projected annual industrial growth rate from 10 per cent to 8 per cent, and industrial investment from 54.7 per cent to 46.8 per cent.[32] The shift in priorities in favour of agriculture, light industry and heavy industry in that order may have had similar dampening effects on the large industrial sector of the provincial economy.

Construction

Notwithstanding a high total investment rate in China, the share of construction investment has been one of the lowest due largely to a very low proportion allocated to residential housing. Until 1977, residential housing has been defined as non-productive and a low priority investment. The share of housing investment ranged between 4 and 9 per cent till 1979 when it exceeded 10 per cent for the first time.[33] However, this share has risen quite substantially since then to reach 25 per cent in 1981. Within the Central–South Region, the share of investment in residential buildings is lowest in Henan and highest in Hunan which also registers more than 50 per cent of the total investment as non-productive (see Table 1.12). Such a high proportion for non-productive construction (public utilities, health centres and residential quarters) indicates the high priority that Hunan province attaches to the material and social well-being of the population.

Construction is generally a labour-intensive activity on account of the low investment requirements. In the case of China, construction employment accounts for 8 per cent of employment in the state-owned enterprises, and 9.9 per cent of the employment in urban collectives. While the share of the Central–South Provinces is either equal to (in the case of Hubei) or below the national average (in the other four provinces) in respect of state enterprises, the situation is different in the case of collective urban employment. The shares of both Guangxi and Hunan in total collective employment are much higher than the figure for China as a whole. This is perhaps due to the preponderance of smaller construction companies in these provinces where construction investment is also concentrated in small projects: over 82 per cent of total investment in Hunan and 73 per cent in Guangxi goes to small projects (see Appendix I, Table A.9).

In addition to urban construction companies, rural communes, brigades and production teams also account for a sizeable amount of construction employment which is not included in Table 1.12. To that extent, the share of construction employment may be underestimated.

TABLE 1.12 *Investment, employment and labour productivity in construction: Central–South China (1981)*

Province	*Share of total investment in (%):*			*Share of construction employment in (%):*		*Labour productivity of state-owned units*	
	Productive construction	*Non-productive construction*		*State-owned units*	*Collective units*	*(yuan/person/ year)*	*Percentage of national average (%)*
		Total	*Residential*				
Henan	69	31	18.2	7.2	9.4	4055	100.0
Hubei	60.1	39.9	25.8	8.2	8.2	4128	101.9
Hunan	47.5	52.5	35.6	6.7	10.2	3512	86.7
Guangxi	55.0	45.0	27.9	7.3	14.6	3294	81.3
Guangdong	56.4	43.6	26.6	4.7	9.1	4511	109.8
All China	58.8	41.2	25.2	8.0	9.9	4051	–

SOURCE *Statistical Yearbook of China for 1981* (People's Republic of China: SSB, 1982) pp. 116–17, 314 and 332.

Average labour productivity in state-owned units in the Central–South Region varies considerably. While labour productivity in Henan and Hubei state-construction enterprises is as high as the national average, that in Hunan and Guangxi enterprises is relatively much lower (Table 1.12). This may be due to the preponderance of the smaller enterprises which tend to be less efficient.

Commerce

At the Fifth Hunan People's Provincial Congress Standing Committee held in May 1980, the Director of the Provincial Commerce Bureau reported an overall increase in goods procurement, stocks and sales. In 1979, as compared with 1978,

> the gross value of procurement of agricultural and sideline products by the provincial commerce system increased by 36.1 per cent, the gross value of procurement of industrial products increased by 17.5 per cent, the gross value of sales increased by 17.9 per cent and the gross value of goods in stock at the end of the year increased by 18.5 per cent.[34]

Further increases have been reported for 1980. It is believed that

TABLE 1.13 *Total value of retail sales in the Central–South Region (1981) (100 million yuan and %)*

Province	*Total value of retail sales*	*(%)*	*Retail sales of consumer goods*	*(%)*	*Retail sales of capital goods for agricultural use*	*(%)*	*Retail sales by peasants to non-agricultural residents*	*(%)*
Henan	135.2	(100)	103.0	(76.2)	28.1	(20.8)	4.1	(3.0)
Hubei	107.9	(100)	85.7	(79.4)	18.0	(16.7)	4.2	(3.9)
Hunan	104.2	(100)	81.1	(77.8)	18.8	(18.1)	4.3	(4.1)
Guangxi	59.1	(100)	44.8	(75.8)	9.2	(15.5)	5.1	(8.7)
Guangdong	176.2	(100)	143.8	(81.7)	22.8	(12.9)	9.6	(5.4)
Central–South Region	582.6	(100)	458.4	(78.7)	96.9	(16.6)	27.3	(4.7)
All China	2 350.0	(100)	1 913.1	(81.4)	347.5	(14.8)	89.4	(3.8)

SOURCE *Statistical Yearbook of China for 1981* (People's Republic of China: SSB, 1982).

these recent increases have been unprecedented for the past decade. In 1981, the total value of retail sales in Hunan was the fourth largest in the Central–South Provinces. The retail sales of consumer goods accounted for the bulk of the total retail sales in all the five provinces including Hunan. These sales of consumer goods seem to be an aggregate of sales to peasants as well as others. It is somewhat surprising to note the peculiar situation in Guangxi where retail sales by the peasants to non-agricultural residents (that is, private retailing) were proportionately higher than in all the other four provinces (see Table 1.13).

System of Ownership

The trends in the evolution of state ownership of commerce and retail distribution since 1950 in Hunan province and in China as a whole are shown in Table 1.14 below. The trends for Changsha (the capital city), Hunan province and China are indeed very similar. Between 1950 and 1975, private retailing had virtually disappeared. Similarly, collectively-owned enterprises had lost importance since most of them were transferred to the state sector. In the mid-1970s, nearly all the shops were reported to be 'owned and run by the State'. Only a few shops were known to be run on a cooperative basis, by a number of small traders and pedlars pooling their savings. Cooperative shops act as retail outlets for the wholesale state shops. In 1976, there were more than 1400 wholesalers and retailers in Changsha city, which belonged to 17 companies handling wholesale distribution of goods. The only trace of private ownership was visible in such service trades as cobblers and barbers.

The process of transition from private to collective and state ownership lasted a number of years. In introducing the joint state–private enterprises, the state undertook to buy shares in the capitalist enterprises. Subsequently the state bought capitalist enterprises completely by paying a fixed interest on the total value of fixed assets of these enterprises. By the eve of the Cultural Revolution in 1965, almost all the state–private enterprises were converted into state enterprises and the practice of interest payments to the private enterprises was discontinued. This led to the replacement of competition and the profit motive by cooperation among stores selling similar kinds of goods. It was not uncommon to find shops with surplus goods helping out other shops which were short of these goods at a particular point in time. Similarly it is reported that state ownership

TABLE 1.14 *System of ownership in retail trade in Hunan and all China (1950–75) (%)*

Year	*State commerce*	*State–private and collectively-owned commerce*	*Private/individual commerce*
Changsha			
1950	11.7	Not recorded	88.3
1956	80.8	19.1	0.1
1975	92.9	7.0*	0.06
Hunan Province			
1950	4.5	0.1	95.4
1956	66.8	27.1	6.1
1975	96.9	3.0*	0.1
All China			
1950	14.9	0.1	85.0
1956	65.7	31.6	2.7
1975	92.5	7.3*	0.2

* After 1966 state–private enterprises were transformed into socialist enterprises. These are figures for collectively-owned commerce.

SOURCE Hsiang Jung and Chin Chi-chu, 'Change of the System of Ownership — Socialist Commerce (I)', *Peking Review* (9 July 1976) no. 28.

helped to promote sharing of transportation and warehousing facilities among different shops and stores.[35]

Rural Trade and Marketing

The structure of the rural trade and marketing network was somewhat different since the supply and marketing cooperatives and trade through rural fairs were still more predominant than state commercial activity. Products not handled by the cooperatives were bought and sold at the regular rural fairs at fixed locations. In the Hunan county of Hanshou in 1976 there were 26 supply and marketing cooperatives with 97 branches controlling 326 purchasing and retail centres.[36] These cooperatives were meant to ensure a steady supply of food and other goods to the urban population and to provide adequate incentive prices to rural producers to increase agricultural production.[37]

After 1978 private trading, discouraged during the early 1970s (see Table 1.14), started expanding rapidly. In 1979, 30 000 rural trade fairs had been restored. A survey of '206 such fairs in 28 provincial level units showed that the total volume of transactions for the fourth quarter of 1978 had increased by 30 per cent over the same period the previous year'.[38] Although most retail shops, hotels, restaurants, repair shops, and so on, in the rural areas are collectively-owned by the rural communes, we were told in Changsha (November 1981) that some rural communes owned restaurants and hotels in the city.[39] These tertiary activities were financed out of the savings of the rural communes which were reinvested in profitable ventures.

Until recently, very little attention in China was paid to trade and marketing activities which were considered as 'unproductive' in a socialist system. Special efforts are now being made to promote these activities as a means of creating additional employment at low capital cost. In the province of Hunan, the Bureau of Commerce has promoted the development of collective commerce which was curtailed in the early 1970s. Many new commercial networks and supply points have been established to serve the consumers better. It is reported, however, that in spite of these initiatives, 'in newly expanding suburban areas and some industrial and mining areas, the masses find it very inconvenient to buy things'.[40]

Pricing of Commodities

According to the commercial department in Hanshou county, Hunan, the average price index for grey cloth, coal, salt, sugar, matches, thermos flasks and ink went down from 100 in 1969 to 85 in 1974, thus indicating a 15 per cent decline. Price reductions in the case of such agricultural inputs as pesticides and chemical fertilisers were as high as 53 per cent. In contrast, the purchasing price of several farm products, for example, wheat, ginned cotton, eggs and fish went up considerably. Table 1.15 shows these price changes in the county for 1957, 1965 and 1975.

During the period covered by Table 1.15, three main types of farm price were in force, namely, basic quota price, above-quota price, and negotiated price. The basic quota prices apply to crops which are sold in fulfilment of procurement quotas, above-quota prices to crops sold in excess of the quota, and negotiated prices to sales by individual farmers and the teams to the state. The figures in Table 1.15 seem to be a weighted average of these three sets of prices.

In 1979, farm price reforms formed a part of the overall economic reforms and readjustments of policies. For China as a whole, in 1979

TABLE 1.15 *Agricultural price changes in Hunan (yuan per kg)*

Year						
Agricultural inputs						
	Chemical fertiliser			*Pesticides*		*Diesel oil*
1957	0.42			1.35		0.26
1965	0.38			1.15		0.16
1974	0.29			1.04		0.16
Farm and sideline products						
	Wheat	*Ginned cotton*	*Jute*	*Pork*	*Fish*	*Eggs*
1957	0.14	1.57	0.74	0.76	0.44	0.80
1965	0.23	1.77	0.84	0.92	0.58	1.24
1975	0.26	2.10	0.84	0.92	0.78	1.32

SOURCE Hsiang Jung and Chin Chi-chu, 'A Vast Rural Market – Socialist Commerce (IV)', *Peking Review* (9 August 1976) nos 32–3.

quota prices were raised by 20.9 per cent for grain, 15.2 per cent for cotton, 25 per cent for oilseeds, and 24.6 per cent for pigs.[41] It is reported that in Zhu Zhou in Hunan province, a rise in the state procurement price for pigs led to peasants crowding at the procurement stations to sell their pigs for fear that the prices might fall again. The sale of pigs in panic is said to have resulted in the death of several pigs.[42] This incident suggests that an appropriate balance between state, collective and private trading and long-term price stability have not yet been achieved either in Hunan or elsewhere in China.[43]

In the price reforms in 1979, a fourth set of prices – free market price – was added with the emergence of private and sideline activities and the freedom allowed to farmers to sell a portion of their produce in open trade fairs and free markets. It is estimated that

> of the total value of net procurement of products of agriculture and sideline industry in 1981, 58 per cent was at quota prices, 21 per cent at negotiated prices, and 9 per cent free market sales to the non-farm population.

In the case of the Central–South Region, the free market retail sales vary between 3 and 8.7 per cent among the five provinces (see Table 1.13).

It is reported that the terms of trade between farm products and industrial goods improved during the 1970s, thus narrowing the 'scissors gap'. By 1978, state procurement prices for farm products rose 11.9 per cent compared to 1965, and the retail prices of industrial goods sold in the countryside dropped by 7.2 per cent. The 'scissors gap' during the period 1965–78 was reduced, according to official estimates, by only 17.1 per cent or 1.5 per cent per year, which is well below the annual rate of 3.6 per cent achieved during the First Five-Year Plan (1953–7).[44]

In the case of Hunan, Riskin maintains that although the prices of such farm inputs as chemical fertilisers, pesticides and diesel oil did go down between 1957 and 1974 (see Table 1.15), 'they were hardly at all in use at the beginning of the period, whereas at the end when badly needed to increase yields, their relative prices were still far higher than elsewhere in the world'.[45] A recent article by Liang Wensen shows the comparative costs of industrial inputs like fertilisers and kerosene in China and the world market as follows: 1 kg of grain can be exchanged for less than 1 kg of fertiliser in China and 2 kg of fertiliser in the world market, and 1 kg of wheat can be exchanged for less than 0.4 kg of kerosene in China and 1.5 kg of kerosene on the world market.[46]

Wiens has estimated marginal domestic prices and border prices (import or export prices adjusted for transport, trade and processing margins to the farm gate level) which show that the marginal domestic price of urea (nitrogenous fertiliser) is 552 yuan per ton compared to its border price of 477 yuan.[47]

The analyses by Riskin and Wiens involving world prices suggest an implicit assumption of a neo-classical framework under which world prices are believed to be competitive and efficient, with openness of economy and free trade possibilities. In reality, these assumptions are unlikely to be true especially in the case of China, notwithstanding the opening up of the Chinese economy in the post-Mao period; neither is the implicit assumption of a strong supply response to price changes on the part of individual farmers, especially under the institutional structure of the Chinese rural economy with production teams as intermediaries between farmers and the state.

CONCLUDING REMARKS

In this chapter, we reviewed economic indicators for the agricultural, industrial and commerce sectors of the provincial economies. Emphasis has been placed on population, employment and income growth particularly during the post-Mao period. It is clear from these data that the Central–South Region of China is characterised by agrarian economies. For instance, this region alone accounted for over 27 per cent of total volume of China's grain output (Table 1.4), and 17.7 per cent of the Chinese gross industrial output (Table 1.8). Increasing emphasis is now placed on industrialisation. In 1981, the region accounted for 27.4 per cent of the Chinese industrial investment in light industry and 17.5 per cent of Chinese investment in heavy industry (Table 1.11). The light industry is relatively more important than heavy industry in the region in respect of output as well. As we shall see in Chapter 6, the region places heavy emphasis on rural industrialisation as an instrument of employment and industrial policy. The commerce and tertiary sectors, which are at present very small, are also to be expanded in order to generate additional employment and to provide additional consumer services for the population.

Thus in one sense, economic transition in the Central–South Region can be interpreted in terms of planned inter-sectoral shifts from agriculture to industry (small and large) and services. However, there is a more fundamental transition from the Mao era to the post-Mao reforms, namely, from a command economy to market socialism, from centralisation to decentralisation of decision-making, and from communes to production teams and households and even individual farmers. It is the transition through these institutional and structural reforms in the post-Mao period in which this book is particularly interested.

2 The New Economic Policy Frame

Since 1978 the Chinese government has adopted a series of measures for economic 'reforms, readjustment and restructuring'.[1] These reforms have been motivated largely by the need to:

(i) ensure greater efficiency in resource use rather than pure resource mobilisation;
(ii) give more emphasis to consumption (and hence to production of consumer goods) than to investment;
(iii) delegate management and planning responsibility to lower echelons of administrative hierarchy with a view to encouraging 'local initiatives';
(iv) provide greater decision-making power to the state enterprises in respect of production-mix and factor use, *and*
(v) promote private sector sideline activities in urban and rural areas for employment and income generation.

The new economic policies are intended to be applied in varying degrees in each province in the light of the local circumstances. Hunan province has started implementing a number of policy measures relating to agriculture, industry and commerce. Since the guiding principles of the new policies are the same for all provinces, it is very difficult to distinguish between the degree of their implementation in Hunan vis-à-vis other provinces in the Central–South Region.

Also it is too early to do any meaningful assessment of the gaps between planned measures and their actual fulfilment. Therefore, in this chapter, we describe and analyse the major policy pronouncements at the provincial and national levels without claiming to assess their outcome fully.

The main reforms which affect the shape and content of the Chinese development strategy and its impact on employment and incomes are: the agricultural pricing policy, the shift in investment priorities from heavy to light industries, the incentives and inducements to commune and brigade enterprises, the household responsibility system to promote private ownership and initiative, and the labour employment system to ensure a more rational manpower utilisation. We shall consider each of these measures below.

AGRICULTURAL PRICING

One of the significant measures adopted is the new pricing policy for agricultural inputs and outputs. We noted in Chapter 1 (Table 1.15) that Hunan province manipulated agricultural prices even in the 1970s to raise farm incomes and encourage agricultural production.

Material incentives to farmers in the form of prices and subsidies are an important feature of Chinese new economic policies in the post-Mao period. An increase in procurement prices since 1979 induced higher farm output and productivity and generated a substantial increase in the average per capita distributed collective income. Manipulation of farm prices has also been used to diversify the rural product-mix. Relatively higher prices for cash crops, for example, have led to a shift in the cultivated area away from foodgrains. Quota prices for cotton were raised by an additional 10 per cent in 1980 when the government realised that an initial increase in price of 1979 induced only a modest rise in cotton production. Although it is not explicit in the Chinese literature, shifts in product-mix could have also been designed to create additional employment through a higher priority for labour-intensive crops.

The price increases were also intended to raise the share of production delivered to the state: a 50 per cent price increase was introduced for above-quota sales to the state for such commodities as grain and cotton which received high priority under the procurement programme.[2]

However, in practice, the growth of agricultural production has not always been rapid in response to price incentives. The costs of agricultural production are known to have increased due to increasingly larger delivery quotas which in turn required the use of high-cost modern inputs. A survey conducted in 1296 production teams in 22 provinces showed that between 1962 and 1976, the average output value per *mou* of six grain crops had increased by 16.61 yuan, which was more than offset by additional production costs of 20.33 yuan.[3] Notwithstanding the decline in farm input prices noted in Table

1.15, the prices for agricultural producers' goods seem to remain high and account for the rising costs in farming.

In many cases, the increase in procurement prices was above the wholesale price of several commodities which meant losses for the state supply and distribution agencies. It is reported that state subsidies to cover those losses amounted to 7.8 billion yuan or 46 per cent of the total budget deficit for 1979.[4] In 1980, the total subsidy arising from increases in procurement prices was of the order of 16.8 billion yuan.[5] For the period 1979–81, the total subsidy on basic foodstuffs and clothing amounted to 41.6 billion yuan. Clearly such large subsidies – involving a transfer of resources from the state to the collective sector – led to a decline in state revenues. The state's attempt to raise the retail prices of some commodities to reduce the enormous subsidies met with resistance from workers who demanded wage increases which in 1980 were estimated to cost 12 billion yuan. The increase in wages in turn is likely to reduce industrial profits and investment.

It is perhaps the decline in state revenues which led the Hunan provincial government to abolish the system of subsidies to some crops. On 23 November 1981, the Hunan People's Government issued a circular on agricultural prices which states that as of

> 1 December 1981, the existing price increases, subsidies and prices that are a combination of assessment and negotiation, for flue-cured tobacco, tong oil and resin will be cancelled, raw lacquer will be controlled as a secondary-category product instead of third-category product. At the same time the regulations to the effect that all localities may increase prices, give subsidies and reduce taxes will be cancelled.[6]

Nor is one allowed to reduce the fixed amount of agricultural products to be procured, enlarge the range of negotiated prices, or increase prices and give subsidies at will. The first-category products are generally considered most important and are therefore under state control. This category originally included 38 articles covering food and vegetables, cash crops and industrial raw materials. The second-category goods are those that are not subject to state monopoly. They originally included 293 articles under the administration of the Ministry of Commerce and the Ministry of Foreign Trade. Any second-category goods not needed by the state are called 'third-category goods' which could be bought and sold in the unregulated markets. The three-tier price system, under which the first- and second-category products are covered by the unified plans of state monopoly purchasing and distribution, as revised in Hunan recently, implies that third-category goods (subject to unregulated trading) would in future not be exchanged freely. Instead many of these goods would be bought by government commercial departments according to signed agreements with communes and brigades who were hitherto free to exchange many goods privately and collectively through trade fairs.

However, it would perhaps be wrong to deduce from the above that handling of third-category goods by the state has been introduced in Hunan on a mass scale. It would be extremely cumbersome to cover diverse and numerous commodities for daily consumption under centralised administration and distribution. Furthermore such a policy of centralisation would also be inconsistent with the new measures relating to decentralised management and the household responsibility system which are discussed below.

INVESTMENT REALLOCATION

A shift in priorities towards light consumer good industries is a very recent phenomenon in China. It is designed primarily to reduce shortages of basic consumption requirements, and, to a lesser extent, to shift production towards greater labour-intensity and employment.

The Chinese authorities believe that the growth in purchasing power will stimulate the demand for the products of light and rural industries (for example, soap, textiles and bicycles), which in turn will induce communes and brigades to expand production within the framework of the new decentralised management of enterprises in the collective sector. Second, the Chinese planners also aim at economising scarce energy resources by restricting the growth of certain branches of heavy industry. In 1978, light industry consumed only 32.2 billion kilowatt-hours of electricity compared to 133.2 billion kilowatt-hours consumed by heavy industry. Electric consumption per unit of output was 0.55 kilowatt-hours per yuan for heavy industry and 0.18 kilowatt-hours for light industry.[7] Third, more rapid expansion of light industry also enables greater employment generation from limited investment resources since the average amount of fixed investment per worker in light industry is estimated at only 6200 yuan compared with 12 000 yuan in heavy industry.[8]

The policy of inter-sectoral investment shifts seems to have been implemented more slowly than was initially expected. For example, the expenditure on capital construction had increased from 30 000 million yuan in 1977 to 45 000 million yuan, representing an investment rate of 36.5 per cent, which was close to the record figure of 39.3 per cent for the Great Leap Forward period. In spite of a plan in early 1979 to reduce the capital investment target from 45 000 million yuan to 36 000 million yuan, the *actual* investments increased from 48 000 million yuan in 1978 to 50 000 million yuan in 1979.[9]

New Measures to Promote Light Industry

In 1979, the provinces of Hubei and Hunan introduced the following measures to promote the development of light industry:

(i) distribution of surplus raw materials: favoured treatment to light industry is given in the allocation of raw material inputs.[10] A basic quota system for commercial agricultural sideline products has been introduced. The surpluses above the quota are distributed between the province and the counties in the proportions of 40:60.

(ii) return of processed products: a system to distribute the profits from increased processing of raw materials has also been introduced. Seventy per cent of the surplus above the plan targets is returned to the localities and counties to promote modernisation and expansion of production.[11] This implies a 30 per cent tax on the local/county above-plan enterprise profits which are collected by the province. This measure seems to have been introduced to prevent localities and counties from indiscriminately establishing small enterprises, particularly in Hunan's tannery industry (see Chapter 6).

(iii) coordination and specialisation: division of labour is being introduced to link supplies of raw materials, production and marketing. Coordinated planning of production from planting of crops to processing of raw materials and manufacture of finished products is being promoted.

(iv) joint operations: since 1979, the light industry department of Hunan has attempted different types of inter-provincial, inter-urban, and interdepartmental companies to overcome shortages of raw materials which tend to hinder the growth of light industry. For example, the Bureau of Shanghai Handicrafts

Industry has negotiated an agreement with porcelain, furniture and paper factories in Hunan and other provinces to supply equipment worth 7 million yuan in exchange for materials needed for its handicrafts production. Similarly, joint operations of enterprises by urban industry and rural communes/brigades can help to utilise excess manpower and idle capital capacity. Sometimes different departments of the provincial administration agree to operate an integrated enterprise jointly, for example, combination of forestry with paper-making, to overcome the problems of raw material supplies for a paper mill.[12]

In the case of joint companies, material supplies between subordinate factories are considered a matter of internal movement rather than a subject for state planned distribution. These internal transfers are not subject to the production tax or profit levy which is imposed on independent enterprises.[13] Thus intra-company costs are reduced.

Notwithstanding the above benefits, the joint companies (or 'transprovincial companies' as they are sometimes called) do not seem to have made much headway in Hunan or elsewhere, due perhaps largely to the problem of equitable sharing of potential economic gains and of the division of responsibility between different parties. Unless there is a strong economic justification for these companies, they are simply likely to add another administrative layer between industrial enterprises and government agencies, thus leading to heavier bureaucracy.

(v) uniting the military and civil sectors: in some provinces like Hubei, defence industries are more developed than civil light industries. It is therefore proposed to boost light industry production with technical support from military industrial plants which are technologically more sophisticated. Sometimes military industries also utilise their surplus capacity to produce light consumer goods for civilian use.

(vi) linking coastal and inland regions: there is wide economic disparity between the eastern (coastal) and western (inland) regions of China. It is proposed that the production of light industry goods currently undertaken in the coastal regions should be gradually transferred to the interior, and to the localities which are near the sources of raw materials. Though not new, this policy measure reiterates that light industry in the inland regions should receive technical assistance from enterprises in the industrially more developed coastal areas. A cooperative establishment of plants by the coastal and inland regions is also proposed. For instance, it is reported that the Shanghai electrical porcelain plant originally shipped pottery clay from the Xianning area of Shanghai for processing in the interior region. Now it is cooperating with the county porcelain manufacturing plant which is producing electrical porcelain objects that are shipped to Shanghai in finished form.[14]

China's new industrial policy has added a number of measures which are designed to favour light industry. In varying degrees, these measures are being applied in Hunan. They include:

(i) allocation of electricity on a preferential basis to enterprises producing light industry goods and export products, and a guarantee of electric power during widespread power failures;
(ii) priority to light industry in transport and supply of scarce raw materials;
(iii) larger imports of raw materials for light industry (for example, cotton and packaging materials) at the expense of imports of steel products for heavy industry.

Heavy Industry *v*. Light Industry

The above policy measures are a response to the failure of heavy industry in serving the needs of agriculture and light industry for simple tools and equipment. The annual plans of Hunan province for 1980 called for sharp declines in different branches of heavy industry.

Heavy industry at present suffers from large excess capacity due to lack of adequate demand for its products. On the other hand, there is a shortage of consumer goods. Therefore as a general industrial policy, it is proposed to utilise surplus capacity in heavy industry to expand light industry output of consumer goods, wherever it is feasible. In order to achieve this objective, a number of measures have been proposed:

(i) provide the light, textile and electronics industry with technical equipment;
(ii) provide the light and textile industries with enough raw and processed materials of improved quality and proper specifications;
(iii) give light industries technical support (presumably through provision of skilled manpower and technical know-how);

(iv) let heavy industrial enterprises produce daily consumer goods requiring technical processes and raw and processed materials similar to those used by the light industry;
(v) transfer to the light industry department some heavy industrial enterprises that have neither long-term or immediate tasks nor any prospects for development, but which are transferable to light industrial goods production.

The policy measures (iv) and (v) concerning the conversion of 'heavy' industry into 'light' consumer goods industry imply, inter alia, malleability of capital equipment. All heavy industry equipment is unlikely to be converted into light consumer goods production without sacrificing efficiency. Switching capital from one industrial use to another depends partly on technical factors like machine specifications. While one of the unique features of Chinese industrial growth (and that of other socialist countries for that matter) is conscious planning to produce multipurpose machines, an absence of standardisation creates problems due to the lack of interchangeability of spare parts.

Plant conversions to remove shortages in essential consumer goods do not always seem to be easy especially in the absence of any systematic planning at the state, provincial and collective levels to match excess capacities of some plants with over-utilisation of the capacity of others which might arise from frequent plant conversions. There is also a danger that conversion of, say, a transport machinery plant into a textile machinery plant might lead to an increased supply of textile machinery but cause shortages of transport equipment.

Currently there is a debate in China on whether the new economic policies have not cut back capital investment and heavy industry too much in order to expand the output of light industry goods. In the long run, this policy might adversely affect technological improvements of the traditional light and rural industries since heavy industry is the major source of machinery required by them. It may also slow down economic growth, and the growth of employment. Yet at the opening session of the National People's Congress held in 1981, Premier Zhao Ziyang stated that the new imbalance (between light and heavy industry) is 'reasonable and proper' in view of the new emphasis on the accelerated production of consumer goods to raise people's living standards.[15]

Contrary to expectations outside China, the output of light industrial goods in 1981 did not rise as fast as predicted. It is reported that it rose by 12 per cent over 1980 instead of the expected rate of 13 per cent, against an expected decline of 8 per cent in heavy industry which in reality was only 5 per cent.

The discussions at the National People's Congress in 1981 suggest that the process of economic readjustment will need to take longer to complete (perhaps five years instead of three) than was originally envisaged.

Nevertheless, China is clearly becoming a more consumer-oriented economy which provides a more favourable economic climate for the growth of rural small industry.

DECENTRALISED ECONOMIC MANAGEMENT

Economic management and entrepreneurial decision-making have been decentralised at two levels, namely (i) at the state-owned, enterprises are now allowed greater initiative and responsibility for decision-making, and (ii) at the collective, brigade and household enterprises outside the state sector are encouraged. Interference from the central and provincial-level authorities, so common before 1977, is now quite limited.

The old relationships between the state and the enterprises have been redefined. Under the old economic policies, it was obligatory on the part of enterprises to fulfil the production targets set by the state which provided fixed and working capital as well as ensuring marketing of their products. Under the new economic regime, on an experimental basis, the enterprises are allowed to produce above the state-determined target and sell the surplus in the open market. They can work out their own production plans in the light of market demand as well as the purchasing plans of the commercial departments. The factories have also been permitted to retain a fixed proportion of profits for technical improvements and employees' welfare. The extent to which the trial-and-error experiments have become an established practice is as yet uncertain, however.

In the collective sector, the historical transition of ownership from the production teams to the brigades and communes which took place during the 1950s is now being reversed. For planning and accounting purposes, the new policies tend to concentrate more on brigades and teams than on communes. The two major features of the recent organisational changes in the commune system are:

(i) abolition of the people's communes as administrative units (the communes will be left with economic and technical activities like

provision of research and extension services, marketing, and so on); *and*
(ii) introduction of the production responsibility system under which contracts are made with individual households, which is a *de facto* recognition of a revival of private farming system (see the following section).

Under the new arrangements, the production brigades and teams operate rural small enterprises relying on their own savings without having to seek investments from the communes.[16] They can also retain most of the profits for reinvestment in capacity expansion and/or establishment of new activities.

In some cases (for example, Jiangsu and Sichuan) communes are being abolished as an experiment since their economic functions are being taken over by the production brigades and teams, and their administrative functions will be performed by the township governments. As of February 1983, Jiangsu has established township governments in all its 68 counties and cities to replace people's communes as the basic organ of administration in the countryside: villagers' committees are also established at the brigade level. In Sichuan, the name 'gongshe' (commune) is no longer used in counties which are trying out new experiments. Instead of the communes, economic organisations called 'agriculture–industry–commerce combines' are established.[17] Although communes have not been abolished in Hunan, probably owing to the Leftist Maoist influence, 23 townships which were abolished during the Cultural Revolution have now been restored to undertake administrative functions hitherto undertaken by the communes.

It is widely recognised that one of the major problems of the communes was the propensity of the cadres to dictate crop distribution and planting methods, to the members thus violating the status of communes as *collectives*.

Nevertheless as economic units, the Chinese communes did indeed perform important productive functions. First, being larger in size than the brigades, the communes enabled economies of scale in production. Second, they provided the efficient management organisation and social infrastructure that are so essential for agricultural and industrial production. Third, a more important benefit of the communes is in terms of non-wage modes of production and employment. Unlike the wage-system prevailing in the state farms, the communes could apply a system of deferred wage payments (in cash and/or kind) and a system of unpaid labour on a large scale. The use of family (non-wage) labour implies that there is very little immediate reward for work thus leading to a low supply price of labour.[18] The reward for greater effort of the commune members would accrue at a later date in the form of greater collective output and incomes. This non-wage system of production is claimed to have resulted in the mobilisation of large masses of Chinese labour for land cultivation and capital construction through water conservation and irrigation without putting pressures on the supply of 'wage goods'. The success of such a system depends on the functioning of moral incentives and group motivation in the communes and the existence of adequate administrative organisation and management. In practice, the experience of the Chinese communes suggests that they might not have been an unmixed success in the use of non-wage modes of labour utilisation without some material, in addition to moral, incentives. For example, it is noted that the supply system of wage-payments in kind, or the policy of 'reward according to need' (under which each family was supposed to receive the same amount of food and clothing regardless of its labour input) had serious disincentive effects. The supply system therefore had to be abandoned in favour of the original wage system.[19] Furthermore, the low supply price of labour of family members may also be illusory under certain circumstances. The exclusive use of work points system in determining reward for labour may have also led to inefficient labour utilisation. Fourth, the communes also played an important role in innovation, experimentation and the diffusion of new agricultural technologies. Their contribution to the rural human capital formation through health and educational/training services cannot be underestimated. Fifth, the communes have also channelled high rates of saving and capital accumulation into the growth of rural industry and the utilisation of labour-intensive technologies.[20] Finally, in the past the communes also acted as the marketing centres for the villages within their areas, and supervised supply and marketing cooperatives.

In the light of the above discussion, it seems doubtful that the policy of disbanding the communes would be consistent with the major objective of employment creation. The abolition of communes may only aggravate the problem of open unemployment and create social unrest.

Group motivation in the Chinese communes partly seems to explain their scale and other advantages. Despite some problems, a combination of 'moral' and 'material' incentives at the commune level seems to have worked well, presumably for cultural reasons. The labour contribution of individual commune members appears to be achieved in the name of collective welfare and goodwill in a number of activities, particularly those of an infrastructural nature.

Ishikawa[21] has argued that, in addition to group motivation, leadership of the party and trust of commune members in the local party and government officials also play a role in determining whether they are motivated to work hard for the collective good.

A gradual shift of responsibility (particularly for rural industrial enterprises, and for planning and investment decisions concerning crops) from communes to production brigades and production teams is basically very similar to the changes which were introduced in the commune organisation in the early 1960s. These organisational units are given greater authority to make their own production and investment plans in the light of local conditions.

A shift to the very small-scale organisational unit implies that in future greater scope should exist for the small-scale labour-intensive enterprises which may employ no more than 8–15 workers. Until recently in China, the 'five small industries' (viz., coal and iron, mining, miniplants for iron and steel, chemical fertilisers and agro-chemicals, cement and machinery), operating at the commune level and employing between 100 and 500 persons, were much larger than what is defined as rural small-scale industry in most other developing countries.

The rural enterprises operated by the production teams are most likely to be of a very small-scale nature, somewhat similar to the rural industries in India. The team usually owns simple tools and equipment and relies on the brigade for larger machinery which is supplied against a service charge. Furthermore constraints of investment resources generated within the team are also likely to limit the scale of production in rural industry. In the province of Hunan, accumulation funds of the production team account for over 8 per cent of the total gross output (sales plus retained production). Table 2.1 indicates how shares of production costs and investment funds have evolved from 1965 until 1979.

Investment funds have to be allocated to activities within the team (for example, rural industry) and to those undertaken at higher levels of ownership, that is, brigades and/or communes.

Under the new economic policies, there is also the household economy, or the private sector, to supplement collective incomes. It is estimated that personal income from private household activities (for

TABLE 2.1 *Distribution of total gross output of Hunan production teams (1965–79) (%)*

Item	Year			
	1965	*1970*	*1975*	*1979*
Production costs	25.4	30.4	33.2	31.9
Taxes	7.2	5.4	4.5	3.5
Accumulation funds	8.2	5.9	9.4	8.7
Distribution to members (as wages based on work points earned)	59.1	58.3	53.0	55.9
Total	100.0	100.0	100.0	100.0

SOURCE Data supplied by the Hunan provincial authorities.

example, animal-raising and private plots) represents 38 per cent of the total rural income compared to 52 per cent from collective activities.[22] This new sector is described in the following section.

PRODUCTION RESPONSIBILITY SYSTEM

Production responsibility system is a generic term used to cover agriculture, industry and other sectors of the Chinese economy. Generally it means that workers are now accountable for specific tasks assigned to them and their remuneration is linked to performance. Under this system, the production teams assign work to each work or production group, link worker remuneration to physical or value output, fix output quotas for households, and allocate time between farming and sideline industrial activities.[23]

As illustrated in Table 2.2, there are a a number of different types of the responsibility system, ranging from production contracted to work groups, households or individuals. The two broad categories noted in Table 2.2 are (i) forms that are not linked to the volume of production (viz. fixed labour quota contracts), and (ii) forms linked to volume of production (for example, specialised work contracts under which workpoints are calculated on the basis of the quantity or value of output). The importance of category (i) has declined from 55.7 per cent in January 1980 to 16.5 per cent of the production teams in October 1981, whereas that of category (ii) has increased sharply from

TABLE 2.2 *Diffusion of different types of production responsibility system (percentage of basic accounting units employing each form)*

Type of system of responsibility	Period			
	Jan. 1980	*Dec. 1980*	*June 1981*	*Oct. 1981*
A. *ding'e baogong* (fixed quota labour contracts)	55.7	39.0	27.2	16.5
B. *zhuanye chengbao* (specialised work contracts)	–	4.7	7.8	5.9
C. *lianchan daozu* (production linked to the group)	24.9	23.6	13.8	10.8
D. *lianchan daolao* (production linked to the worker)	3.1	8.6	14.4	15.8
E. *bufen baochan daohu* (production partially contracted to household)	0.026	0.5	–	3.7
F. *baochan daohu* (production contracted to household)	1.0	9.4	16.9	7.1
G. *baogan daohu* (operations contracted to household)	0.02	5.0	11.3	38.0
H. Subtotal for linked forms (B to G)	29.0	51.8	64.2	81.3
I. Total for all forms of the system	84.7	90.8	91.4	97.8

SOURCE *Jingjixue Zhoubao* (*Economic Studies Weekly*), 11 January 1982, quoted in Katsuhiko Hama, 'China's Agricultural Production Responsibility System', *China Newsletter*, JETRO, Tokyo, no. 40, September–October 1982, p.3.

29 per cent to 81.3 per cent during the same period (see Table 2.2).[24]

The specialised production or work groups are given full responsibility for certain types of output, subject to a target negotiated by a contract between a production team and a group. The labour remuneration is determined on the basis of the fulfilment of the target. In the case of *baochan daohu*, the individual households are required to produce and hand over to the team a certain stipulated amount of output. Any amount of output attained in excess of the contracted amount can be retained and sold by the households. This case is very similar to the fixed-rent tenancy under capitalist farming. The contract between the team and the household does not affect the collective ownership of land and major farm tools and equipment. Only the actual farm work is divided among smaller units. It appears that the *baochan daohu* is yielding place to *baogan daohu* under which specific operations like use and maintenance of farm implements is entrusted to households. These two variants of the responsibility system are prevalent in Guangdong, Guangxi and Henan in the Central–South Region (see Table 2.3).[25] In the case of Henan and Hubei, contracts linking production volume to the individual worker (*lianchan daolao*) are prevalent.

Within the Central–South Region, the adoption of the responsibility system seems to be less marked in provinces of Hubei and Hunan. For example, in 1981 the Hunan authorities had declared that they would not permit any contracts to individual households in the province.[26] It is plausible that the Hunan party officials are still followers of Mao and Hua who are reluctant to adopt measures that are being experimented with in Deng's province of Sichuan. For example, there are press reports of Hunan party officials and government functionaries resisting the introduction or widespread diffusion of the production responsibility system. It is stated that 'officials confiscated trucks and tractors that some farmers had bought to carry produce to market arguing that this sideline was capitalistic'.[27] Domes has also reported opposition to the responsibility system in several provinces, notably Hunan, Hubei and Guangxi, in contrast to a strong support for it in the province of Sichuan.

The agricultural responsibility system is somewhat different from the cultivation of private plots. The former is applied to 'collective' fields (even where these are now in fact cultivated by individual farmers) while the latter refers to land set aside for household use, all the output of which is privately used.

In the case of private plots also, Hunan and Hubei were, for a long time, less advanced than several other provinces. Towards the end of 1979, while the province of Sichuan introduced a ceiling of 15 per cent for the share of private plots in the overall cultivable land, Hubei and Hunan kept the share at 5–7 per cent at least till the end of 1981. In the case of Hunan, Domes argues that the share of private plots was

TABLE 2.3 *Adoption of the production responsibility system by form and province (percentage of production teams as of June 1981)*

Specialised work contracts (zhuanye chengbao)		*Production linked to individual worker (lianchan daolao)*		*Production linked to the group (lianchan daozu)*		*Production and operations linked to household (baochan daohu and baogan daohu)*	
Hunan	18	*Henan*	53.3	Beijing	53.2	Guizhou	9.5
Guangxi	15	*Hubei*	21.4	Tibet	49.2	Gansu	72.2
Heilongjiang	45	Hebei	46.9	Jiangsu	33.4	Anhui	69.3
Jilin	41.4	Shandong	30.3	Xianjiang	31.6	Ningxia	57.9
Shaanxi	17.4	Shanxi	25.4	Shaanxi	28.0	*Guangdong*	41.8
Shanxi	15	Jiangsu	16.8	Jiangxi	25.4	Inner Mongolia	40.1
		Beijing	16.8	Liaoning	25.3	Tibet	39.2
		Shaanxi	16.4	Sichuan	25.1	*Guangxi*	35.7
				Fujian	20.7	Yunnan	35.5
						Xinjiang	33.3
						Henan	33.2
						Fujian	33.1
						Jiangxi	under 30
						Sichuan	under 30

SOURCE *Bulletin of China's Economic Research Organisations* (*Quanguo Jingji Tuanti Tongxun*), 10 August 1981 (no. 18, 1981). Cited in Katsuhiko Hama, 'China's Agricultural Production Responsibility System'.

kept low because 'close associates of Hua-Guo-feng are still in charge'.[28] By early 1983, in most provinces including Hunan (and presumably Hubei), the share of private plots had expanded to between 12 and 15 per cent, whereas in some counties of Sichuan this share had reached as high as 25 per cent.

It is too early to make a proper assessment of the agricultural responsibility system which has not yet been fully implemented in all the provinces. However, the Hunan provincial authorities claim that the communes and brigades covered by the responsibility system have higher outputs and incomes than those which are still outside the system. For example, in the Wangling commune, the increase in average output per team adopting the system during 1979–80 was estimated at 290 *jin* per *mou*. Similarly the Bazishao commune of Yiyang county applying the system registered an increase of 122 per cent in grain production between 1978 and 1980. Further, in Wanyu commune (consisting of 217 production teams and 40 000 people) in Huarong county, the fixed assets at the commune, brigade and team levels have increased since the application of the responsibility system. This is shown in Table 2.4 below. Each accounting unit contracted production and distribution to individual households who retained the output over and above that stipulated in the contract. The fixed assets may have been acquired with the sale proceeds of such retained output. It is, however, difficult to ascertain whether all the claimed increases can be attributed exclusively to the responsibility system.

The responsibility system may have its own problems too. For example, the system of contracting to individuals may lead to income inequality and reduction in capital accumulation. Peasant households and individual workers with more fertile and larger plots of land are likely to benefit more than those with smaller and less fertile plots. Similarly, households with a limited manpower will be handicapped compared with those having a larger number of members. It is reported that the introduction of the responsibility system has led to a decline in school enrolment since households with a shortage of labour tend to put school-age children to work at home. It is further recognised that 'individual management is not conducive to population control. The increase of the rate of population growth in 1981 to 1.4% up from the previous year's rate of 1.2% is considered to have been stimulated by the diffusion of the production responsibility system.'[29] The new system is also likely to blur the earlier distinctions between capitalist and socialist forms of organisation in agriculture. Opposition to the new system in Hunan province seems to be

TABLE 2.4 *Increase of fixed assets in Wanyu commune (Hunan) under the rural responsibility system*

Fixed assets	*1978*	*1982*	*Percentage increase (%)*
Owned by the production teams			
Hand tractors (no.)	34	77	126.5
Water buffaloes (no.)	2 896	3 125	7.9
Electric motors (no.)	148	278	87.8
Power sprayers (no.)	0	106	–
Total fixed assets (000 yuan)	3 050	3 900	27.8
Owned by the production brigades			
Large tractors (no.)	12	12	0
Diesel engines (no.)	51	95	86.2
Total fixed assets (000 yuan)	2 260	3 090	36.7
Owned by the commune			
Large tractors (no.)	3	3	0
Electric motors (no.)	15	15	0
Motor vehicles (no.)	0	3	–
Total fixed assets (000 yuan)	429	500	16.5

SOURCE Peng Xianchu, 'Rural Responsibility System: Spot Report (on Hunan), Part I', *China Reconstructs*, vol. XXXI (August 1982) no. 8, p. 55.

symptomatic of some apprehensions along these lines. In the absence of the commune structure, difficulties may also be encountered in the provision of public services and agricultural input supplies, and management/maintenance of (farm) machinery.

Nevertheless it is reasonable to assume that the transfer of responsibility to teams and households under the responsibility system offers material incentives which may partly account for higher output and productivity. Collectivisation of agriculture during the 1950s and 1960s may have been carried far beyond the scale and management capacity of the Chinese peasants. This may be particularly true of backward areas or regions with sparse and dispersed farm population.[30]

Although the production responsibility system was first introduced to farming and animal husbandry, some of its features seem also to have penetrated commune and brigade enterprises engaged in handicrafts and other sideline productions, marketing and sale of products.

There is a counterpart of the new agricultural responsibility system in the commune and brigade industries in the sense that the small industries formerly belonging to the communes are now being owned and operated by the production teams, households and even individuals.

Outside agriculture, the promotion of private sideline activities is likely to encourage the growth of seasonal rural industry (of an artisanal type) which can be operated by women and young and old men during agricultural slacks. Previously households were expected to allocate a minimum number of labour days in a year to collective production, which limited the supply of labour for sideline activity. However, recent rules for collective labour are more liberal: only one working member from each household is required to contribute to collective labour. Although the amount of labour available is still determined by the size of the family and its composition, one or more members can now wholly withdraw from collective production to undertake sideline activity. Both working and non-working household members can undertake private activity.

The fact that the capital base of the household is more limited than that of the production team further suggests that village/cottage crafts industries prevalent in countries like India are perhaps the most suited for the Chinese private activities. There are also indications of the growth of such non-industrial activities as rabbit breeding, bee-keeping, the picking of wild medicinal herbs, and so on, in response to such incentives as cheap bank loans and marketing assistance by the government extension agencies and marketing cooperatives.

Private sideline activities are designed to increase rural employment and the supply of certain consumer goods, the demand for which is increasing with an increase in the personal disposable incomes. It is estimated that nearly 30 per cent of the rural labour force is surplus and could be mobilised through rural non-farm activities.

Although a number of restrictions on the private sideline activities have been lifted, three broad guidelines govern their nature and magnitude, namely: (i) sideline production should not damage national resources, (ii) it should not affect collective production adversely, and (iii) it should not lead to private trading and speculation. It appears that the Hunan provincial authorities are particularly concerned that the growth of the private economy might take place at the expense of the collective sector. The households are therefore not allowed to hire outside labour which restricts the supply of labour for sideline activities particularly in those households with only a few working members. Second, a number of rural small enterprises have been merged into larger ones to ensure efficiency and profitable production (see Chapter 6). This implies that these larger enterprises are operated more efficiently on the larger scale of communes and brigades than at the level of the individual or the household. It is therefore doubtful whether the disappearance of the communes, and possibly brigades, would be without any economic and social costs.

The responsibility system is also being applied to the industrial sector, though perhaps to a lesser extent at present. It consists mainly of such measures as profit retention (*lirun liucheng*), profit and loss guarantees (*yingkui baogan*), and income-tax payments instead of transfer of profits to the state by the industrial enterprises. There are two main types of this system, namely:

(i) the enterprise responsibility system regulating rights and obligations between the state and the enterprises, *and*
(ii) the intra-enterprise system which coordinates relations between departments of an industrial enterprise and between its workers and staff.[31]

In the case of type (i), the enterprises can now retain most of the after-tax profits. It is worth noting, however, that the 'collective enterprises' in the 'second line of light industry' were already paying income-tax instead of transferring profits to the state. Under the new system, the individual enterprise, and not the relevant department, is responsible for its profits and losses. The type (ii) system encompasses a number of rewards to labour through workpoints, piece-rate wage payments for production in excess of quotas, and rewards through contracts for specialised assignments undertaken by offices, workshops, factories, and even individual workers.

One of the major advantages of the economic responsibility system in the industrial sector is that it provides a material incentive for the enterprises for profitable production, and an additional source of finance for modernisation and reinvestment. In principle, the system should encourage rapid industrial growth through reinvestment of surplus by the enterprises. However, one possible problem is that the state might lose an important source of revenue. In a planned socialist economy like the Chinese, the state is more likely to be able to reinvest surplus to achieve such social objectives as employment creation than the individual industrial enterprises. In the case of divergence between private and social objectives, the new system might well promote industrial growth without social equity and employment.

On an experimental basis, the responsibility system is also being extended to commercial and service establishments. In January 1983 a pilot project on 'responsibility system in management' was launched in Beijing in about 400 state and collective shops, department stores and service establishments covering such activities as general merchandise, garments, groceries and catering and repair shops. Various forms are being tried depending on the size and management characteristics of the establishments. For example, the smaller shops just pay income-tax and assume control over their profits and losses. The larger shops sign contracts guaranteeing the state a fixed sum and keep half the profits they earn over that sum. The catering and service establishments pay the state 20 per cent of their profits as income tax, retaining the rest. It is expected that similar pilot projects will be implemented in other parts of the country.[32]

LABOUR EMPLOYMENT SYSTEM

Prior to 1977, the state assigned all jobs through central and local labour bureaus assisted (in the case of college graduates) by the Ministry of Education and the State Planning Commission. The labour departments/bureaus in different provinces allocated jobs and workers to different factories and enterprises. Workers could not be hired or fired, nor could the factories promote or demote them in response to their good or poor performance. In the case of the technical and vocational school graduates, the ministries or enterprises running these schools had the responsibility for labour allocation. An annual labour plan, to be approved by the planning agencies at the central, provincial and county levels, determined the disposition of the new entrants to the labour force among different enterprises, specified the amount of migration to be allowed from the communes to the urban areas, and the number of youths to be settled in the communes. This system of the state assigning jobs to individuals and guaranteeing employment has now undergone considerable change.

After the Third Plenary Session of the Eleventh Party Central Committee held in 1978, voluntary collective employment and individual self-employment were combined with job allocation by the state within the framework of a unified plan. The National People's Congress held in Beijing in September 1980 enunciated a new employment policy with the following notable features:

(i) allocation of jobs to college and secondary technical school-leavers (employed by the state enterprises) under a unified state plan. If enough skilled workers are not available for the state sector, the best qualified job seekers also become eligible; the remainder are encouraged to work in collective and individual enterprises;
(ii) encouragement of enterprises that combine farming, industry and commerce. These are mainly farms linked to processing plants, workshops and retail outlets to be staffed largely by secondary-school graduates;
(iii) shorter working hours and more shifts in factories to provide additional jobs;
(iv) greater assistance to cooperatives and one-man businesses through special loans and exemptions from income-tax for the first three years of operation;
(v) provision of technical training for the unemployed.

One of the main features of the Chinese current employment policy is the promotion of self-employment of individuals in trade and commerce and other service activities (for example, restaurants, noodle shops and workshops for repair and maintenance of machinery) or in what is called the 'informal sector' in Western literature.[33] This individual economy is seen as an important supplement to the state and collective sectors. Urban collective enterprises are another source of employment accounting for nearly 10 per cent of net industrial output. In 1980, employment in the urban collectives increased by 7 per cent or nearly twice the rate of increase of employment in state enterprises. This was to be expected since it is estimated that the 'fixed assets per employee in state–owned enterprises' are 10 000 yuan whereas those in collective enterprises are only 2000 yuan.[34] The Hunan Labour Bureau recently reformed the employment system so that the 'university graduates, technical school-leavers, and demobilised soldiers' are encouraged to work in urban collectives.

It is also planned in Hunan and elsewhere to give preference to the absorption of educated youth seeking jobs. In 1983, Hunan Labour Bureau has instructed the state enterprises and collectives above the county level to make special efforts to employ youth graduating from training and vocational schools. 'Youth farms' specialising in profitable cash crops and linked to agricultural-industrial-commercial corporations, are being set up to find employment for urban youth in rural areas. *Xingdan* in Hunan is one such model farm.[35] Most of these farms are developed specially for the absorption of 'rusticated' urban youth. During the Mao period, one of the major policies for controlling the size of China's urban population was to deport youth from the cities to 'the mountains and down to the countryside' to settle on the collective farms. In 1979, official estimates indicated that 1 million of the 3 million urban middle-school students graduating annually would be settled on these collective farms.[36] These transfers are claimed to have provided skills in the rural areas which the farmers lacked. Apart from the development of the countryside, rustication of urban youth was also intended to reduce the urban population pressure and provide ideological re-education.[37]

The recent policy of decollectivisation and abolition of communes is likely to add a new dimension to the rural employment problem. A large number of commune officials are expected to become redundant in the wake of the radical rural reforms. It is not quite clear how they are to be reabsorbed. The dynamic and competent ones may take over supervisory functions in the new production cooperatives and corporations. However, a report by a study group of the State Agricultural Commission found that many commune officials were too incompetent and uneducated to be suitable for new positions. It is reported that some of them are being transferred to local birth control and welfare programmes.[38] The extent to which these officials compete with the rusticated urban youth for given jobs in rural areas is not known.

Technical training for the unemployed job seekers is provided through the urban labour service companies which, though abolished in the 1960s with the elimination of unemployment, have re-emerged since 1977. These service companies are set up by the labour departments to organise vocational training courses, support individual establishments and supply casual and contract labour to enterprises. To date, more than 11 000 labour service companies have been established in 140 of China's 197 large and medium-sized cities, and 600–700 of over 2000 counties. These companies have set up 58 000 production and service networks and have assisted in the creation of 1.26 million employment opportunities. Furthermore they have organised 1.61 million people as casual labourers and labour teams, and trained 320 000 people.[39]

Until recently the enterprises did not have any clearcut recruitment and personnel policies with the result that labour skills were allocated to them on an *ad hoc* basis. Such a policy did not satisfactorily meet the requirements of industry for skilled manpower. None of the factories that we visited was allowed to hire workers in the open market. In the Cable and Wire Factory in Xiantan (which was one of the most modern visited by us) the technical staff was allocated by the Hunan provincial government in Changsha. This and other factories visited could not lay off staff or transfer them to other factories within the province or outside. The management of many complained of serious shortages of technicians. We also found few examples of any systematic assessments of workers' proficiency except awards in the form of badges and medals to individual workers and teams for above-average productivity. The rigid system of job allocation by the provincial labour bureau meant job security for workers regardless of their performance. This hindered the improvement of labour productivity, and in some cases led to situations of over-staffing and mismatch between jobs and skills, and between the supply of and demand for labour.

The new employment policy seeks to recruit workers by a method of applications and examination on an experimental basis in a small number of factories in Hunan. For example, in August 1981, over a thousand urban unemployed youth went to Zhu Zhou ramie textile mill to take part in an examination organised by the mill to recruit 200 workers. The Zhu Zhou Municipal Labour Department selected candidates purely on the basis of their performance.[40] This new system is intended to give better options to job seekers and to match the right skills with the job vancancies.

The Hunan Labour Bureau has also decided to abolish the old system of the 'iron rice bowl and of everybody eating from the same big pot' which implied job security and equal distribution of wages and bonuses. All the factories that we visited in November 1981 hired labour on the basis of a fixed monthly wage. None of them practised the workpoint system which was known to have been employed in rural collective enterprises. The new policy implies that, in future, the fixed-wage system will be replaced by flexible and piece-rate wages linked to productivity, output and skills.[41]

As a result of the new policy, Hunan province has witnessed four main changes in the employment pattern of its labour force. First, major commercial and service trades rather than industry are now claimed to account for the bulk of employment. Second, collective and household economic activities rather than the state-run enterprises are said to have become a major source of employment. However, it is doubtful that outside agriculture, household activity yet provides a significant share of total employment. Of course, the collective sector does, but then it did even before the economic reforms were introduced in Hunan and elsewhere. Third, job seekers waiting for employment in small towns will in future be absorbed locally. Formerly, people seeking jobs in urban areas were sent back to the countryside. The Chinese policy-makers believe that helping the job seekers find a 'livelihood by themselves' in commerce and services will facilitate job creation. Finally, the educated youth are to be employed in farms and factories in the city suburbs and towns instead of being 'rusticated' to the countryside as was the case during the Cultural Revolution.[42]

Currently, the control of labour movement from the rural areas to the towns through administrative methods prevents the urban employment problem from growing worse. The old policy of recruiting urban workers from the countryside has also been discontinued to ease the urban unemployment situation. Thus the new liberal employment policy cannot lead to rural-to-urban migration (and convert disguised unemployment in the rural areas into open unemployment in the urban areas) as is indeed the case in many other developing countries. However, just like other developing countries, China plans to create rapid employment opportunities within rural areas through rural diversification via the promotion of non-farm activities.

Since late 1982, state enterprises have also been allowed to hire workers on a contract basis for a fixed period.[43] This measure is very similar to the production responsibility system applied in agriculture and industry under which contractual arrangements are made to get a specified amount and type of work to be done. The state enterprises can enter into contracts with workers, each contract defining the terms and conditions of wage payments, fringe benefits, conditions of termination of employment, and so on. Workers are required to hand over a portion (10 yuan) every month to their neighbourhood labour service companies as a contribution to a social insurance fund. A survey of 160 000 contract workers in nine provinces, municipalities and autonomous regions which have introduced the contract employment system showed that the new measure was beneficial to both workers and enterprises. Under the new system, workers tend to work harder to make their jobs secure and to earn additional income. The enterprises are free to dismiss workers whose performance is unsatisfactory.

3 The Technology Policy Frame

The rapid growth of employment through the state, collective and private enterprises envisaged by the Chinese planners will depend, *inter alia*, on the nature of the technology used and the product-mix. For example, it is estimated that in 1978 for every

> million yuans' investment in heavy industry, jobs can be provided for 94 persons, the same investment in light industry can generate jobs for 257 persons; in the garment trade, arts and crafts and metalware for daily use, and leather and leather products, as many as 800 persons can be employed.[1]

It is therefore appropriate to examine shifts in Chinese technology policy over time especially since the fall of the 'Gang of Four'.

The most recent call for a technological transformation of China is embodied in the policy of 'four modernisations' of which technology is one along with agriculture, industry and defence. The task of all-embracing modernisation is a gigantic one which is concerned, *inter alia*, with technology choice and investment allocation, with the process of technology generation and adaptation (through formal and informal R and D), with the process of technological diffusion within the economy, and with the import of advanced technology from abroad. In general, the emphasis on modernisation should involve a higher degree of capital-intensity and mechanisation and a lower rate of employment expansion. The simultaneous attainment of the two objectives of rapid modernisation and employment expansion therefore poses a real challenge to the Chinese planners.

In this chapter, the Chinese technology policies are examined in a historical perspective. For this purpose, the period 1949–82 is divided into four phases culminating in the post-Mao period.

TECHNOLOGY CHOICE AND INVESTMENT ALLOCATION

Little is explicit about the precise ways in which investment planning and technical choice decisions have been made in China in the past. In general, these decisions involve the determination of the rate of investment, and the pattern of its allocation in different sectors, industrial branches, products, projects, and so on. Non-Chinese scholars like Eckstein and Ishikawa have speculated about the Chinese system of resource allocation and project planning. Eckstein reported that new industrial construction is often divided into 'above-norm' and 'below-norm' projects. Above-norm projects are those which require an investment ranging from 10 to 30 million yuan, and are therefore under central control and planning. Projects requiring less than 10 million yuan of investment are considered 'below-norm'; the planning of these is presumably decentralised to the provinces, prefectures and counties.[2] Ishikawa has given a more rigorous explanation of the Chinese criteria of investment allocation. He noted that starting with the Great Leap period, the Chinese economy was divided into 'central' and 'local' sectors (perhaps as a kind of planned dualism). The centrally-controlled investment resources were allocated between the two sectors in the light of a growth maximisation criterion in such a way as 'to equalise the marginal productivities of the central funds in the respective sectors'.[3] During the 1960s and the early 1970s when technology imports were allowed, there emerged another allocational problem, viz. that of an optimal allocation of funds between technology imports and indigenous R and D effort.

It is not clear whether the central and provincial planning commissions use any explicit criteria for the choice of an optimum combination of technology imports and indigenous technologies. Neither is it clear whether a conscious planning effort is made to choose an optimum combination of techniques from known indigenous alternatives. However, in the Chinese literature, there are references to the 'economic effectiveness' of investments which seems to be used as a major criterion for project selection. More recently, it has been supplemented by employment creation as an additional criterion. The recent introduction of a new discipline entitled 'technological economics' also suggests an increasing consciousness on the part of the Chinese planners for the use of systematic economic and social criteria (something equivalent to the social cost–benefit analysis) in project and technology selection. Wang Enkui defines the main task of this new discipline as 'to measure, compare, analyse and evaluate the economic results of technological measures and programmes adopted in construction and production as well as for the implementation of technology policy'.[4]

The Chinese criterion of economic effectiveness seems to be no different from the 'coefficient of investment effectiveness' or its inverse, the 'period of recoupment' used in the Soviet Union.[5] According to Ishikawa,[6] this criterion was used in China during the Soviet period (1950–7) and, with some modification, in the subsequent periods as well. For example, in the Great Leap Forward phase, the coefficient was revised upwards and the recoupment period downwards since the Chinese planners put greater emphasis on the smaller-scale and quicker-yielding investment projects. This modification of the coefficient, which implies choice of lower capital-intensity, need not be inconsistent with the objective of maximising the growth rate which guided both Soviet and Chinese planners. As Dobb has pointed out, in such a situation the amount of surplus per unit of capital may be smaller but the surplus could be reinvested earlier given the shorter gestation period, thus leading to a 'compounding effect'.

We consider the following four time periods to trace the technology evolution in China: the Soviet phase (1950–7), the Great Leap Forward (1958–60), the Cultural Revolution (1966–76) and the post-Mao phase (1976–82).[7] Variations in the degree of capital-intensity (a crude indicator of technology) during these phases can be illustrated as in Figure 3.1. On the vertical axis is measured capital, K, and on the horizontal axis labour, L. The slope of the different vectors originating from O indicates the degree of capital-intensity whereas the *amount* of capital allocated to each technique is given by the height of the points on the given vector. In the diagram, the points *S, P, M* and *R* are on the same horizontal line $K'K''$ which implies that the amount of capital allocated to large-scale capital-intensive technique ('leg one') was the same in each period. The reality, however, need not correspond to this situation: relative shifts of investment – from industry to agriculture, and within industry, between heavy and light segments – which occurred during these phases must have entailed shifts in the amount of capital invested in the capital-intensive (leg-one) and labour-intensive (leg-two) techniques.

Soviet Phase

During the Soviet phase, the degree of capital-intensity, at least in industry, was quite high. This is shown by the vector OPN in Figure 3.1. If all capital were invested in the Russian technique LL_1 amount of labour would have remained redundant. Since China was a full-employment economy at that time with every member of the work-force being guaranteed employment by the state, a residual amount of capital had to be employed with a large amount of labour thus resulting in a very low capital–labour ratio (K/L) as indicated by the ray *QP*. Full employment was ensured by simply leaving the bulk of the annual net additions to the labour force to remain underemployed in agriculture and in labour-intensive handicrafts.

The Soviet phase coincided mainly with the Chinese First Five-Year Plan (1953–7) which marked the period of rapid industrialisation through investment in heavy industry. A high priority for heavy industry at the expense of agriculture seems to explain a relatively high incremental capital–output ratio for this period estimated at 2.0 compared to 0.8 for 1950–3 and 1.3 for 1958–60.[8] The emphasis on

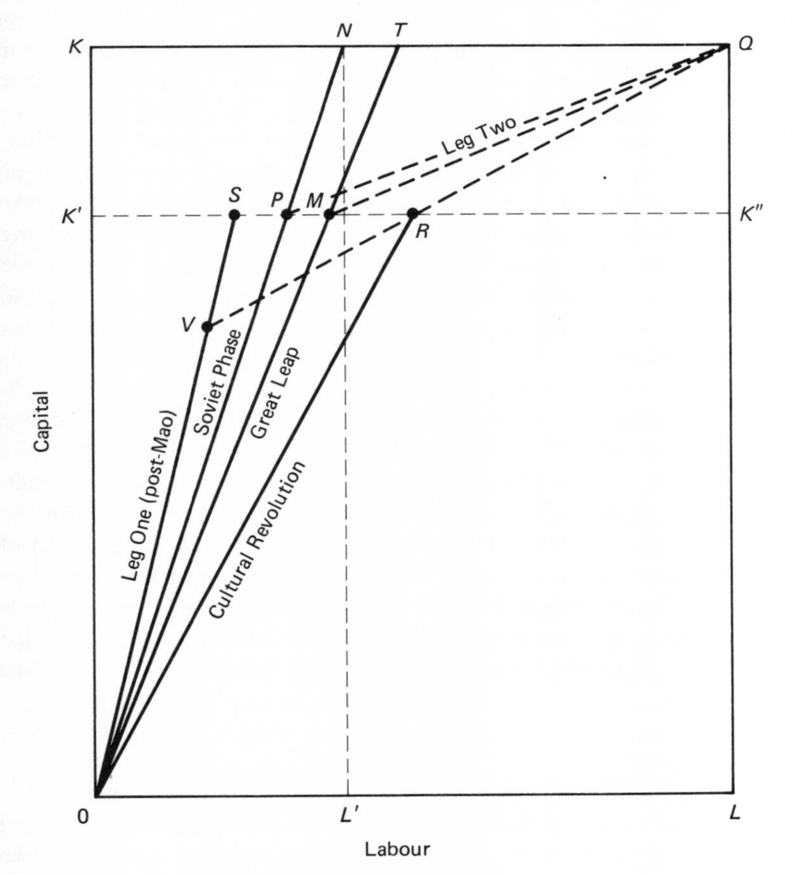

FIGURE 3.1 *Technology Variations in China*

NOTE Rays indicating leg one and leg two in the diagram are only illustrative of the four phases. They indicate average capital-intensity in the economy and not in the industrial sector alone.

such heavy industries as steel, engineering, machinery and petroleum, and gradual mechanisation also tended to raise the capital–labour ratios during the Soviet phase.[9] It would appear that K/L ratios increased further during this period, at least for industry as a whole. Plans for industrial growth consisted mainly of a group of about 150 large modern projects built with Soviet aid. These projects accounted for 11 000 million yuan of investment out of a total of 25 030 million yuan during 1953–7. The shares of Henan, Hubei and Hunan were, respectively, 654 million, 802 million and 350 million yuan.[10]

According to Genevieve Dean, to maximise employment by using

> 'less mechanised' labour-intensive technologies implied limited or no technological advance for small and medium-scale enterprises. No reference was made to the development of labour-intensive technologies implying the use of existing technologies: obsolete, modern as well as traditional. The restriction on technological advance in these enterprises was enforced through the state monopoly on investment resources and capital goods representing 'advanced' technology. In effect, by not specifying alternative sources for new small-scale, labour-intensive technologies, the planners failed to implement any technology policy at all in this sector.[11]

The Great Leap

This period roughly coincided with the departure of the Russian technicians along with their blueprints. Russian technology was no longer available, and even if it were, the Chinese management and organisational capacity was too limited to utilise imported technology. As a result, the Chinese leaders favoured a self-reliant strategy based on technological dualism through a 'walking on two legs' policy. This planned technological dualism may at first appear to be an inconsistent strategy. Yet there can be an economic rationale for this, especially under non-constant returns to scale, as is shown by Sen and others.

The Great Leap period witnessed mass mobilisation of labour for capital construction *à la* Nurkse through highly labour-intensive and large-scale projects. For the first time, extremely labour-intensive techniques were used to produce output which was hitherto produced only by the capital-intensive techniques. In the industrial sector it meant the coexistence of 'leg-one' and 'leg-two' techniques. In terms of Figure 3.1, the leg-one technique is represented by OMT, and the leg–two technique by line QM. During this period, the share of investment in agriculture also increased whereas that in industry (both light and heavy) declined somewhat.[12] To the extent that agriculture represents smaller-scale, more labour-intensive projects, such a shift in industry-mix should have lowered the overall degree of capital-intensity. Greater emphasis was also placed on the completion of unfinished projects and their maintenance. Preference was given to the fuller utilisation of existing capacity than to installation of new capacity. All these factors seemed to have lowered the capital-intensity of leg-one projects in this period relative to that in the Soviet period. This is shown by the flatter slope of OMT than of OPN in Figure 3.1. It is particularly difficult to determine capital-intensity for the short period of the Great Leap. The leg-one projects completed during this period were very capital-intensive as in the Soviet phase since both were Soviet-aid projects. However, new projects *begun* during this period must have been much less capital-intensive. One could therefore assume that the average capital-intensity of leg-one projects in both periods was quite similar, although leg-two projects were really started during the Great Leap. Technical improvements in leg-two techniques and projects of a very labour-intensive nature (indicated by the ray QM in the diagram) are another notable feature of this period. To quote an example of improvements in construction techniques, cable-towed scoops and wagons running on wooden rails were introduced to raise the efficiency of digging and earth-moving.[13] Similarly, simple multi-purpose machine tools were developed for small-scale industry.

An interdependence between the two legs is a peculiar feature of China's technological policy without a good parallel in any other developing country. Ishikawa has analysed this relationship in terms of a concept of 'investment inducement mechanism'. In the case of small 'leg-two' projects, local resources with zero or low opportunity cost are the major source of investment. However, some central resources were provided to induce the mobilisation of these local resources. The heavy emphasis on leg-two projects generated a demand for equipment which was supplied by the relatively large-scale machine-goods sector. The Chinese experience shows that productivity improvements in the small-scale sector depend partly on the rate at which indigenous innovations occur in the large-scale sector – a

TABLE 3.1 *Capital–labour and gross output–labour ratios in Chinese state industrial enterprises (yuan)*

Year	*Capital–labour ratio*		*Gross output–labour ratio*
1952	3 157*	3 113†	4 167
1957	4 582*	4 518†	6 336
1965	9 090*	9 063†	8 943
1975	9 512*	10 064†	9 994
1979	11 485*	11 554†	11 790

* Refers to original purchase value of fixed capital per worker (including staff).

† Refers to total capital (original purchase value of fixed capital net of depreciation plus working capital determined by certain norms).

SOURCE *ZGJJNJ 1981 (Almanac of Chinese Economy)*. Taken from S. Ishikawa, 'China's Economic Growth in the PRC Period — An Assessment', *China Quarterly*, (June 1983, no. 94).

factor which has been often overlooked in most developing countries.

Cultural Revolution

It is reported that the level of technology in the manufacturing sector advanced considerably during 1964–7 due presumably to the importation of 'turnkey' projects from the Western countries. Table 3.1 gives estimates for capital–labour and output–labour ratios for state industrial enterprises (which are usually large-scale leg-one types) for a number of years. It shows that the size of the capital–labour ratio more than doubled from 1957 (the eve of the Great Leap) to 1965 (the eve of the Cultural Revolution). However, between 1965 and 1975, the increase in K/L ratio was not so significant which may be explained by a number of factors, namely, fuller utilisation of existing equipment capacity and revival of urban small-scale enterprises requiring low investment per worker.

The technical improvements of leg-two labour-intensive projects started during the Great Leap seem to have been more successful during the early 1960s when they had time to be tried and diffused widely. The 'walking on two legs strategy' continued during this period in a more structured and rationalised manner not only within industry but also within industrial enterprises. Wheelwright and McFarlane, who visited China during the Cultural Revolution, describe their trip to the Xinhua Printing Works in Changsha (Hunan) in the following words:

> there is a wide variation in machines which mould type. Machines with different levels of technology and sophistication are all working simultaneously. Four of the machines were mechanised to a fair degree and were made in Shanghai in 1957; four others work by hand and pedal. An old machine is also operating but mainly 'for education' as it can do only 40 pounds of type mould per eight–hour shift, while the Shanghai ones can do 170 pounds per shift.[14]

The leg-two techniques (QR) during this period were far more capital-intensive than those during the Great Leap, due mainly to gradual mechanisation of rural small–scale industry.

Post-Mao Period

During the initial phase of this period, considerable emphasis was placed on the import of most modern foreign technology for implementing the 'four modernisations' policy. This phenomenon, which is sometimes described as the 'foreign leap forward of 1977–8' may partly explain an increase in capital–labour ratios in 1979 as noted in Table 3.1.

However, the Chinese authorities soon realised that they were rushing too far ahead of their capacity to pay for and absorb imported technology. After 1980, the technology imports were cut down partly on account of the scarcity of foreign exchange and partly to protect a Chinese infant capital goods sector. Imports of machinery which can be produced domestically would henceforth be discouraged. Similarly, agricultural mechanisation is being slowed down (see Chapter 5), whereas in industry the new policy is to utilise fully the existing productive capacity rather than expand it. Furthermore the value of investment in heavy industry had already started declining from 1976 onwards. It is estimated that it declined from 92 058 million yuan during the Fourth Plan period (1971–5) to 65 783 million yuan in 1976–8. During the same periods, investment in farm machinery, chemical fertilisers and pesticides went down from 9 338 million yuan to 7316 million yuan.[15] There was a further decline in the share of investment in heavy industry from 77.3 per cent in 1978 to 74.7 per cent in 1979.

Table 3.2 below gives investment–employment ratios by sector for

TABLE 3.2 *Investment–employment ratios by sector in state-owned units: Central–South Provinces (1981) (yuan)*

Province	*Ratios*					
	Total economy	*Agriculture*	*Industry*	*Construction*	*Transport*	*Commerce*
Henan	446.0	914.6	668.8	108.8	370.7	180.7
Hubei	499.5	210.0	818.0	130.4	293.6	232.4
Hunan	375.0	228.8	382.0	129.4	328.3	334.0
Guangxi	316.6	402.5	415.4	69.6	221.5	164.0
Guangdong	628.1	574.8	663.8	117.1	1 133.9	397.0
Central–South Region	617.4	435.7	619.0	115.0	544.5	281.7
All China	511.0	391.7	631.9	137.6	805.7	244.7

NOTE The coefficient for agriculture may be somewhat misleading since the denominator consists of the number of workers in state farms only whereas the bulk of employment is accounted for by rural collectives.

SOURCE *Statistical Yearbook of China for 1981* (People's Republic of China: SSB, 1982). The above figures are derived from the table on 'Number of Staff and Workers in State-owned Units By Province' on p. 116, and the table on 'Investment By Sector of National Economy and Province' on pp. 310–12.

1981 for the Central-South Provinces. The wide variations in these ratios across sectors and provinces suggest that inter-sectoral investment shifts can be manipulated to generate greater employment from given total investment in state-owned production units in different sectors.

Another factor that may explain the slowing down of increasing capital-intensity in industry during this period is the effort 'to separate, or decompose, advanced technologies so that some operations can be carried out with labour-intensive techniques without affecting product quality'.[16] This was particularly the case with the use of manual techniques in such ancillary operations as materials-handling and transport in the Hunan factories that we visited. For example, in a spark plugs factory the packaging of spark plugs into cartons was done manually with very high speed. Although the factory management thought of mechanisation of this operation, it did not really seem to be necessary for raising productivity which could just as well be achieved through better work organisation and flow (see Appendix II).

One of the main characteristics of the post-Mao period is a marked shift in emphasis from capital accumulation and rapid economic growth to consumption by the masses. This is reflected in the doubling of annual growth of private consumption from 2.7 per cent during 1960-70 to 5.4 per cent in 1970-80. During these two periods, the annual rate of growth of gross domestic investment went down from 9.8 per cent to 6.8 per cent.[17]

The Chinese sources[18] report that the old problem of excessive accumulation has yielded place to that of overspending caused by the growth of average wages relative to output and productivity growth, redistribution of an increasing portion of the national income to the peasants through prices, taxes and bonuses, social welfare, and subsidies and so on for agricultural and sideline products. To quote, 'the consumption funds have increased by more than 96 billion yuan in the past three years (1979-81) which is 10 billion yuan more than the increase in national income in the same period'. In spite of this shift towards consumption, the current savings rate in China (30 per cent in 1980 in contrast to 23 per cent in 1960) is one of the highest in the developing world. It is also higher than the savings rates of the most advanced market and socialist economies. Thus one cannot easily assume any sub-optimality of savings during the post-Mao period which might have been the basis of Chinese planning during the earlier periods, particularly the 1950s and early 1960s. In China today, greater emphasis on consumption than on investment implies that technology is no longer used as an instrument to raise savings and capital accumulation. Instead it is more likely to be used to generate employment, incomes and consumption.

RESEARCH, DEVELOPMENT AND MANPOWER

Technological self-reliance and national capacity to import and assimilate foreign technology presuppose the existence of adequate research and development facilities, availability of scientific and skilled manpower and well-developed administrative organisation for R and D and technology diffusion. These are the issues to which we now turn.

The issue of R and D and of its linkage to production has been subject to variations over the four time periods examined above. A major difference between the Soviet period and the subsequent periods seems to be in respect of the organisation – centralisation or decentralisation – of research. In the 1950s, the bulk of R and D must have been concentrated on the problems of heavy industry and of the adaptation and utilisation of imported technology. With the decentralisation of planning and the introduction of the 'walking on two legs policy' during the Great Leap period, R and D was also decentralised and a number of local research institutes established.[19] In addition, a shift towards light industry is also likely to have led to a reorientation of R and D priorities in favour of domestic instead of imported technology and small-scale instead of large-scale sectors.

Two-legs Strategy in R and D

The Chinese policy of 'walking on two legs' can also be applied to R and D. The advanced R and D undertaken in academy and industrial institutes, and required for innovations in the modern industrial sector and for the adaptation and assimilation of imported technology into it roughly corresponds to one leg. 'Mass science' and 'Mass innovation' undertaken by 'extension groups' at the commune and brigade levels in the Chinese rural economy represents the second leg. However, as we shall discuss below, the second leg involves more adaptation and redesign than basic and applied research for which the rural economy has to depend on 'leg-one' R and D institutes. The technology used in the rural sector originates mainly from outside: only adjustments to acquired technology are made at the local level.

The relationship between the two legs in R and D is established through a vertical hierarchical link: the research institutes at the central or provincial/regional level supply mechanical and biological technology to the county enterprises; the county institutes supply new designs, equipment, technical guidance and training, seeds, fertilisers and so on to communes; communes offer technical advice and rental services of farm machinery to production brigades and teams. This vertical hierarchy in R and D is but one manifestation of the vertical economic management in China which may discourage horizontal communication within each leg. Acquisition of technology, research funds and raw materials are regulated through administrative orders from central ministries, provincial bureaus or local authorities at the county and commune levels.[20]

Linking Research and Production

China's example is often quoted as a successful case of links between R and D in the central institutes and actual production. There is evidence that the practice of 'going out' (by research teams from institutes to production enterprises to do on-the-job R and D) and 'inviting in' (of workers and technicians from factories into institutes to influence research designs) has led to some integration of research with the productive process. However, recent Chinese literature suggests that this objective has not yet been fully achieved. It is reported that only 14 per cent of the applied research results of Academia Sinica's institutes in 1977 were actually popularised.[21] The Hunan Provincial Meeting on Science and Technology held in May 1981 also noted that scientific research in the province was not linked to actual production.

The above problem seems to arise from the administrative and planning system in China. The major industrial enterprises design and manufacture equipment on the basis of plan targets and administrative orders. In the past, there was little opportunity for the users of the machinery to influence its design. Thus equipment manufacturers failed to get any feedback either at the design stage or after its installation, when invariably the repair and maintenance is done by the users and not by the equipment producers and suppliers.[22]

The Hunan Meeting on Science and Technology also noted that scientific research in the province was not linked to actual production. It further realised that the plan targets for science and technology for the province were unrealistic. In the light of its conclusions, the meeting recommended:

(i) linking of science and technology with economic development;
(ii) proper analysis of alternative production techniques;
(iii) development and promotion of new technology in industrial and mining enterprises where its application is known to be slow; *and*
(iv) application of foreign advanced technology wherever appropriate.[23]

The conclusions and recommendations of China's first Conference on Rural Scientific Research and Application (February 1983) suggest that the rural sector has not benefited all that much from vertical technology transfers from higher levels. The conference discussed the role of scientific developments and innovations in agricultural growth. The agricultural responsibility system leading gradually to private farming is likely to create additional demand for scientific inputs and local technology systems. It is planned, therefore, to raise output of medium- and low-yield plots which at present account for 78 per cent of China's arable land.

The rural conference considered a proposal to establish a national agricultural technology development centre which is intended to provide an institutional framework for the application of science and technology for rural development. Experimental bases have already been set up under the joint sponsorship of the Chinese Academy of Sciences and the provincial governments to conduct intensive surveys of natural resources, and to organise agricultural scientists to conduct experiments in research centres. Each base is formed with a county boundary as its boundary. The following five counties have initially been chosen: Taoyuan in Hunan, Luanzheng in Hebei, Hailun in Heilongjiang, and Guyuan and Yanchi in Ningxia.

The 'three-in-one' groups so popular during the Mao period were also mainly engaged in agricultural innovations. They were ill-equipped to do any applied industrial research. Furthermore the resources of the communes for R and D are either very limited or non-existent. Whatever little R and D takes place relates not to any mechanical innovations but instead to agriculture/crop research and experimentation. The agricultural research and extension section at the level of communes consists of technical workers, electricians, mechanics, truck drivers, and so on. Very few of these personnel are engaged in R and D related tasks. They are usually junior-middle school graduates with some on-the-job training in farm implements factories and repair workshops.

Shopfloor Innovations

Apart from research in the industrial institutes, many large industrial enterprises also conduct some in-house research. But we got the impression from our field visits that the bulk of this work related not so much to R and D as to simple design and adaptations of products and processes. Most of the factories visited reported that generally 2–3 per cent of their total expenditures was allocated to R and D work. In order to get an impression of innovations at the enterprise level, we give below a few case study illustrations of our factory visits in Hunan.

Spark Plugs Factory

The porcelain department of the factory produced porcelain parts with manually operated machines which seemed to operate quite fast. However, the R and D department of the factory has developed a prototype of an automatic machine which is designed to replace the manually operated machines. This automatic machine will displace manual labour almost entirely. This was one of the few factories whose R and D department had actually attempted to introduce innovations. Apart from the R and D work, the factory has its own technical school to train workers and raise their skill levels. Training courses cover both technical and cultural topics. The R and D department has 75 engineers and technicians on its staff. Technicians are particularly scarce. The department does not do any basic R and D but is a testing centre for experimentation and applied research. All the products are designed and manufactured by the plant itself without any external assistance. In addition to the products made according to the standards set by the National Standards body, the plant can also design and manufacture spark plugs of all sorts through consultations with, and according to, specifications of the clients from abroad. The R and D department plays an important role in improving the quality of plant's products and processes.

Cable and Wire Factory

The factory has four R and D departments working on technology adaptation and on new products. Currently work is in progress on different kinds of alloys to produce wires and cables more cheaply. In some cases the factory has shifted from the use of copper (which is scarce) to that of aluminium. The factory expressed the need for the following technology which is currently not available in China: (i) fissions of solid physics; (ii) experimental and testing equipment; and (iii) precision and measurement instruments.

Diesel Engine Factory

The factory is currently engaged in developing new models which will be lighter (substituting aluminium for cast iron), with greater horsepower and lower fuel consumption. In the development of these new products the factory engineers (in its research department) were assisted by a research institute in Shanghai. The latter provided technical assistance: testing and experimentation of the prototypes was done in the institute. The factory has a unique contractual arrangement with the Shanghai research institute with the result that the latter cannot provide similar technical assistance to other factories.

Not all the industrial enterprises that we visited had R and D departments. It would perhaps be safe to assume that R and D within industry at present forms only a very small fraction of total R and D, at least in the province of Hunan.

One reason why this situation had prevailed until recently seems to lie in the centralised industrial structure. Recent policies and reforms of industrial reorganisation and decentralisation of economic management of enterprises, under which the enterprises can now retain a large portion of after-tax profits, should in future facilitate in-house innovations and research. Such a development is likely to reduce the dependence of enterprises on higher-level research organs. It is too early to indicate whether such a trend will reduce vertical linkages and, instead, promote horizontal linkages among enterprises.[24]

A recent UNCTAD survey of 47 leading Chinese enterprises manufacturing machine tools and heavy electrical equipment shows that all these enterprises did in-house R and D related to the development of new products.[25] These new products, defined as 'those whose technical performance had been remarkably improved', were introduced in the sample firms at the rate of 9.3 products per firm during the 1975–80 period. The actual R and D and product development are shared between the enterprises and the research institutes, the former concentrating presumably on the product adaptation and manufacture. It is noted that the Ministry of Machine Building responsible for the sample firms allocated a very low R and D allowance which was supplemented by the firms' own funds to the tune of between 50 and 300 per cent of the Ministry allowance. It is worth noting, however, that these enterprises are some of the most advanced ones in the capital-goods sector. Their experience is unlikely to be valid for the smaller enterprises in the consumer-goods sector.

R and D Expenditure

Hardly any systematic statistics exist on total R and D expenditures for China. However, Sigurdson and Billgren made a bold attempt (based on certain heroic assumptions) to estimate R and D expenditures by major categories for 1973. They estimated a total figure of 4.59 billion yuan for 1973 which was allocated as follows: 0.11 billion yuan for basic research (2.4 per cent of the total), 0.81 billion yuan for agriculture and natural resources (18 per cent), 0.52 billion yuan for medicine or public health (11 per cent), 1 billion yuan for defence (22 per cent) and 2.15 billion yuan (47 per cent) for manufacturing, energy and transportation.[26] In a recent study Suttmeier analyses these Billgren–Sigurdson estimates and finds them 'too conservative' even though they represented about 1 per cent of China's GNP in 1973.[27]

The Chinese government has released a figure of 5.87 billion yuan for national R and D expenditure for 1979, which represents a 10 per cent increase over 1978. In the light of this figure, it appears that the Billgren–Sigurdson estimates for 1973 (that is, six years earlier) might have been reasonable. The Chinese estimates of R and D expenditure cover investment in laboratories and other facilities, the wage bill for scientists and researchers, and the costs of specific pilot projects. Of the total R and D budget, 585 million (10 per cent) is allocated to basic research, 1.17 billion (20 per cent) to applied research and 4.1 billion yuan (7 per cent) to development work.[28] If we take the Billgren–Sigurdson figures as realistic, the estimates for 1979 would suggest a considerable increase in the importance given to basic research, rising from 2.4 per cent of the total budget in 1973 to 10 per cent in 1979.

Two main factors seem to account for the increase in total R and D budget, namely, the creation of new research institutes which entailed expansion in capital construction, and an increase in the wage bill resulting from increase in the number of scientific personnel, promotions and salaries.[29]

TABLE 3.3 *Investment in scientific research: Central–South Provinces (1981)*

Province	*Total amount (100 million yuan)*	*Percentage of total for China*	*Percentage of total for the Region*
Henan	0.31	3.3	23.6
Hubei	0.41	4.2	31.0
Hunan	0.10	1.0	7.6
Guangxi	0.12	1.3	9.0
Guangdong	0.38	4.0	28.8
Central–South Region	1.32	14.1	100.0
All China	9.34	100.0	

SOURCE *Statistical Yearbook of China for 1981* (People's Republic of China: SSB, 1982) p. 312.

More recent data on investment in scientific research in the provinces of the Central–South Region is given in Table 3.3 below. It shows that the share of Hunan is the lowest, followed by Guangxi. The shares of Guangdong and Hubei, the two more industrial provinces in the region, are much higher.

Starting 1 January 1982, the Chinese Academy of Sciences (CAS) has established a Science Fund to support efforts to tap the country's scientific potential, train scientists, strengthen basic research and promote scientific undertakings. Scientific workers in all departments are eligible for a grant of funds for research. The Fund is intended mainly for basic research in natural and applied sciences. It is largely financed by national resources although donations from within China and abroad are also accepted.[30]

Scientific Manpower

According to the Chinese official estimates, in 1981 there were about 5 714 000 natural scientists and technical personnel engaged in state-owned undertakings. Of these, 36.4 per cent were engineering staff, 29.4 per cent were in public health and medicine, 22.6 per cent in teaching, 5.9 per cent in scientific research and 5.7 per cent in agriculture (see Table 3.4). Data on the Central–South Provinces shows that the Guangxi autonomous region, which is known to be poorer than the other provinces, has a smaller share of different types of

TABLE 3.4 *Natural scientific and technical personnel in state-owned units: Central–South Provinces (1981) (000)*

Province	*Total manpower*		*By category*									
			Engineering		*Agriculture*		*Public health*		*Scientific research*		*Teaching*	
Henan	260.7	(21.5)	78.3	(19.9)	15.4	(19.0)	86.6	(22.7)	9.4	(18.0)	70.9	(23.2)
Hubei	316.8	(26.1)	100.6	(25.6)	15.6	(19.2)	100.9	(26.5)	15.3	(29.5)	84.1	(27.5)
Hunan	241.0	(19.9)	80.3	(20.4)	16.7	(20.6)	70.9	(18.6)	9.4	(18.0)	63.5	(20.8)
Guangxi	160.8	(13.2)	54.7	(13.9)	14.6	(18.0)	52.4	(13.8)	6.0	(11.6)	33.0	(10.8)
Guangdong	233.9	(19.3)	79.4	(20.2)	18.8	(23.2)	69.8	(18.4)	11.9	(22.9)	54.0	(17.7)
Central–South Region	1 213.2	(100)	393.3	(100)	81.1	(100)	380.6	(100)	52.0	(100)	305.5	(100)
All China	5 713.9	(100)	2 076.8	(36.4)	328.1	(5.7)	1 680.2	(29.4)	337.5	(5.9)	1 291.2	(22.6)

NOTE Figures in parentheses indicate percentages.

SOURCE *Statistical Yearbook of China for 1981* (People's Republic of China: SSB, 1982) p. 464.

scientific personnel in the total for the whole region. In the case of Hunan, the share of engineering manpower is somewhat lower than that of Hubei although it is somewhat higher than that of Guangdong. This is rather surprising since this province is known to be much more industrial than Hunan. It is plausible that Hunan accounts for a larger proportion of agricultural engineers who may be included in this category. Hunan's share of scientific research personnel is quite low which is to be expected since we noted in Table 3.3 that its share in investment in scientific research is also quite small. In general, Henan, Hubei and Hunan have a greater concentration of scientific and technical manpower than either Guangdong or Guangxi.

The industrial enterprises which we visited in Hunan (many of which were state-owned) all complained of a severe shortage of middle-level manpower like technicians. The shortage of engineers to operate foreign machinery and plants also seemed to be quite serious. A number of Hunan industrial enterprises send their technical staff to Western Europe and North America for advanced training under trade and know-how agreements in order to overcome these shortages. For example, a piston rings factory visited by us had a contractual agreement with a British firm to train its staff in the UK, and to impart in-factory training in Changsha (see Appendix II).

However, recent press reports from China indicate new restrictions on Chinese students going abroad for higher studies and training. These restrictions are presumably intended to prevent too much exposure to Western cultural influences and consumption patterns.[31]

The shortages of scientists and technicians are felt most acutely in the rural areas. It is reported that only 320 000 such personnel work in rural China, that is, less than 0.04 per cent of China's rural population.[32] In order to overcome these shortages, the provinces like Hunan send peasants to attend agricultural colleges. In Hunan, there are 70 000 such colleges financed by public welfare funds. Most of the teachers for these institutions are drawn from research institutes and farm machinery bureaus.[33]

Part-time scientific and technical education is also rapidly developing in the rural areas of Hunan. At the end of 1980, more than 23 000 part-time technical schools were set up with a total enrolment of more than half a million students, 194 000 short-term training classes, and over 841 000 attendance of commune members.[34]

The Chinese central government is quite conscious of the need to encourage more scientists and engineers to work in rural areas. It plans to introduce such material incentives as 'floating wages' and 'awards of appropriate professional titles' for inducing the transfer of scientific manpower from the urban to rural areas.

TECHNOLOGY DIFFUSION

The rate of diffusion of innovations seems to have varied during the four time phases considered above. One can argue that internal diffusion accelerated during the periods of self-reliance (for instance, the Great Leap and part of the Cultural Revolution) and may have slowed down during the periods of external dependence through technology imports (for instance, the Soviet phase and post-Mao era). In the former case, national self-reliance and reduction of imports meant domestic manufacture of equipment and components, learning-by-doing and replication and diffusion of known innovations throughout China. This process might have been hindered when liberal imports of technology and equipment meant their localisation in a small segment of modern 'heavy' industry. The reports that, between 1972 and 1979, 252 duplicate sets of equipment covering 17 categories of items were imported, suggest that information flows within the Chinese economy were hampered, presumably by a lack of specialisation resulting from the vertical hierarchical structure of industry and bureaucracy. Baark, for example, quotes a slow diffusion process within the Chinese petroleum industry and attributes it to 'the management system prevailing in the Chinese scientific and technological research as well as the management of the Chinese enterprises generally.[35]

Nevertheless, China does have a well-established institutional structure for technology diffusion. A number of provincial, county and commune-level institutions are engaged in disseminating technological information to the end-users, besides a Central Institute of Scientific and Technical Information. It is reported that the Departments of State Council and various provinces, municipalities and autonomous regions have established 72 scientific and technological information research institutes. In addition to these, there are about 3000 national and regional specialised scientific and technological information centres or stations. Over 50 000 full-time technical staff are engaged in these institutes and centres. At present, more than 370 scientific and technical journals are published throughout China.[36] The Central Institute of Scientific and Technical Information regularly translates and publishes translations of foreign literature on science and technology and distributes them to the provincial institutions.

A conference, organised by the State Scientific and Technological Commission in 1980 in Beijing, emphasised the importance of broadening the sources of information, collection and documentation of technical data, and keeping abreast of scientific and technological achievements and trends in China and abroad.

The province of Hunan, like most other Chinese provinces, has a number of technical information centres at different levels of administrative hierarchy. For example, at the provincial level, there is an information centre on farm machinery which falls under the Bureau of Machine-Building Industry. It ensures the exchange of information through the publication of nearly 200 journals including an English-language journal entitled '*Agricultural Machinery Digest*'. Furthermore, the Hunan Farm Machinery Research Institute in Changsha (see Appendix IV) also has a division on technical information and standardisation which serves as one of the sources of technical reference material and blueprints. Third, the production and technology department of the Hunan Bureau of Commune and Brigade Enterprises has an information network dealing with cement and paper-making, the two important rural industries in the province. Fourth, the Bureau of Commune and Brigade Enterprises has a separate information department which distributes technical information to rural enterprises.

Technological information is also distributed to commune and brigade enterprises through:

(i) training courses lasting from two weeks to three months which are held either at the communes or at the relevant educational institutions;
(ii) the transfer of technical personnel (engineers, technicians) from the production departments of rural industry to the communes and brigades; *and*
(iii) the supply of leaflets/brochures on new technological innovations.

Agricultural research institutions and teaching/training colleges in Hunan also perform technology diffusion functions besides undertaking research and training. For example, the agricultural technicians supplied by the county R and D institutes to the communes and brigades not only organise scientific experiments and training courses but also promote links with other agricultural institutions in order to ensure inter-commune diffusion of innovations. In addition, the 'May 7 Peasant Colleges' located at the county, commune and brigade levels, also participate in the dissemination of technical information through training courses of varying duration.[37] These colleges are usually small and undertake short courses of two weeks to three months on such subjects as plant protection, insect control and animal husbandry.

The local technology diffusion system in Hunan and elsewhere in China appears to be far more advanced for agriculture than for rural industry. The information dissemination at the county, commune and brigade levels relates mainly to agricultural production. For example, evening classes held at the communes and brigades mainly discuss the use of seeds and their selection, fertiliser application and water conservancy. Similarly the 'personal demonstration' method under which experienced peasants are asked to transfer their knowledge to their colleagues is also confined to farming. Close links exist as well between peasant colleges, secondary agricultural technical schools and provincial agricultural colleges. Such links and channels of communication seem to be much less developed for non-agricultural activities.[38]

The reasons for such an imbalance are not quite clear. It is plausible that the crop failures in the early 1960s and subsequent high priority for agriculture led the Chinese government to put a relatively greater emphasis on agricultural research. This is likely to be particularly true in the case of Hunan which is predominantly an agricultural province.

Writings of Western scholars on China, particularly those of Sigurdson, suggest a highly developed local technology system in China, consisting of mass scientific networks, local research institutes, technology transfer, diffusion mechanisms, and so on.[39] Contrary to our expectations, we were told that, at least in Hunan, no such formal local system existed at the commune and brigade levels specifically for the benefit of rural industry, although such networks were common between larger enterprises of the same kind, such as cement or machine tool fabrication. Two reasons seem to explain this situation. First, there is a lack of resources at the level of the brigades and communes for undertaking any major tasks relating to research, design and development and diffusion of technology. Second, the productive capacity in rural areas is highly diversified which makes the exchange of information at a horizontal level extremely difficult.[40]

To overcome some of these problems, the state now urges all the scientific and educational institutions to train rural industry personnel and to supply technological information suitable for rural industry. Notwithstanding these efforts and the existing institutions, the Chinese participants at the Changsha International Seminar on the Modernisation of Industry related to Agriculture (November 1981), felt a need for a centralised formal information network. Such a network would initially concentrate on a few priority areas and economic sectors. It would embrace central, provincial, county, commune and brigade levels of hierarchy. The recommendations of the Changsha Seminar include the establishment of a Hunan Technical Information Centre for rural small industry 'to develop, on the basis of existing institutions, a formal network for the collection and dissemination of information to rural industrial enterprises and responsible government organisations'.[41] This further reinforces our argument that the local technology systems in Hunan are not systematically developed.

Within the industrial sector, in general, administrative, management and organisational bottlenecks tend to hinder technology diffusion. Apart from administrative barriers which prevent inter-enterprise communication, the decentralisation of decision-making is also said to have led to what is described by one author as 'technological blockades'.[42] Particular industrial enterprises which possess specialised technology (often imported from abroad) are reluctant to share it with others especially in conditions of increasing competition. This problem is well described in the following quotation:

> Technological blockades actually exist in our country. After expanding the autonomy of enterprises, competition arose, and secrecy of technological processes became an ominous matter. To gain prize money, or for reasons of competition, the inventors would often lock up an invention or the content of the new technology, and even at the appraisal meetings sponsored by higher level organs, when the substance of inventions is introduced, the inventor would refrain from divulging key points of the technology. This phenomenon of blockages is becoming a serious trend. If this trend is not corrected, it will adversely affect the development of new technologies and create a state of stagnation.[43]

Lack of linkages between the industrial sector and the research institutes of the Academy of Sciences is also a problem which has been recognised by the Academy. A number of measures are being considered to overcome this problem; one of the most notable ones is that an increasing share of funds for the R and D institutes should come from industrial contracts and fees.

Technology transfers in China of a functional (between large and small enterprises) and geographical (between rural and urban areas and between coastal and inland regions) nature are aimed at improving the backward rural areas through rural industrialisation. It is therefore not surprising that in the Chinese literature one comes across much more discussion of rural and urban linkages than of internal linkages within rural areas. The process of technology diffusion is characterised by vertical links between large and small industry much more than the horizontal exchanges of know-how and information. The examples of horizontal flows are much less prevalent simply because, at given scales of enterprises, the differences between levels of productivity and technology are less varied than at different scales of operation in a vertical hierarchy.

ADMINISTRATIVE ORGANISATION OF SCIENCE AND TECHNOLOGY

Our four periods were marked by the relative neglect of formal science and technology plans and policies until the post-Mao period, and revival since then of the science and technology institutions which were closed during the Cultural Revolution. The revival of science and technology commissions and associations at the central and provincial levels, formulation of a long-term science and technology plan for the 1978–85 period, and the development of a legal framework for the utilisation of foreign technology, investment through joint ventures, compensation trade and licensing and know-how agreements, are a few examples of this new structure.

At the highest national level, the State Science and Technology Commission cooperates with the State Economic Commission and the State Planning Commission in formulating and implementing national science and technology plans. Below this level are the ministries which translate broad policy directives from the Commission into concrete projects on basic and applied research. At the third level, the ministerial activities are decentralised at the levels of provinces which have bureaus falling under a particular central ministry. At the fourth level are the enterprises which undertake productive activity and applied research.

For the province of Hunan, the administrative structure/hierarchy of science and technology institutions as re-established after the Cultural Revolution is described in Figure 3.2. At the apex there is a provincial committee of science and technology and a science and technology association. While the Committee (on a par with a bureau) is a political body with executive authority to approve technology transfers between provinces and down to lower levels of administration, the Association is a scientific body to promote professional exchanges and organise seminars. Under the provincial Academy of Sciences, a number of science and technology research institutes exist, relating to agriculture, forestry, fishery, livestock and animal husbandry, and sideline production. These institutes deal with basic scientific research, for example an institute for agriculture would deal with high-yielding varieties, food engineering, and so on. On the other hand, the Agricultural Machinery Research Institutes fall under the Bureau of Machine Industry and deal mainly with mechanical technology. They undertake research by projects and promote information dissemination.

At the level of regions or districts also similar research institutes exist which deal with the same topics as those covered by the provincial institutes. Below the county level there are no institutes but only Science and Technology Research Groups. No such groups exist at the level of production teams. There are only 'extension groups' – a sort of a 'mass scientific movement' – to popularise new ideas and apply them as well as to provide skills to artisans. Local R and D, diffusion of information, and improvement of machinery are generally beyond the capacity of these extension groups. This 'mass scientific movement' in China seems to be similar to rural extension services in other developing countries with one possible difference, however. In the case of China, the extension movement has been supported by popularisation stations, experimental farms and local research institutes, thus enabling a close link between R and D and production activity in the rural areas. The numbers of different institutes and science and technology groups in Hunan are given in Table 3.5.

In addition to the data in Table 3.5, there are also a number of secondary agricultural technical schools located in the surrounding prefectures and attached to the Hunan Provincial Agricultural College which is engaged in teaching and research. The graduates of the Agricultural College are assigned by the provincial government to work at the provincial, prefecture and county levels. These graduates are

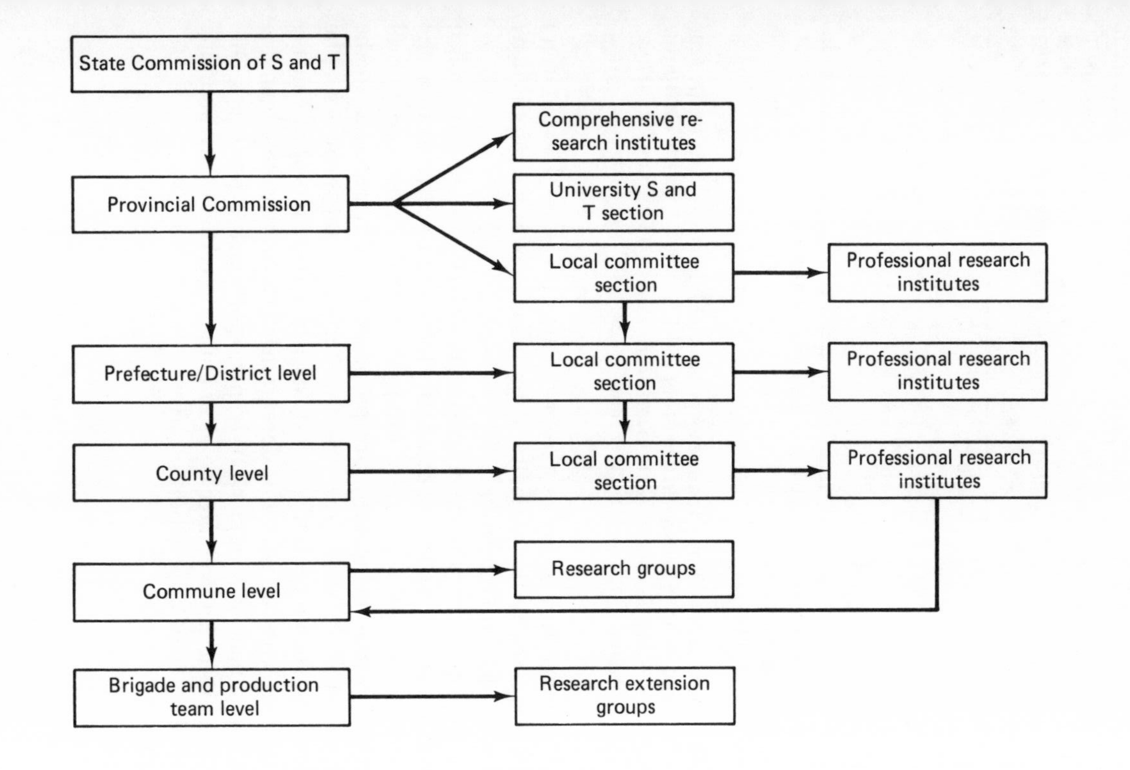

FIGURE 3.2 *Administrative structure of science and technology (S and T) in Hunan*

TABLE 3.5 *Science and technology institutes in Hunan (no.)*

Administrative level		*Scientific research institutes*	*Institutes of science and engineering*	*Institutes of agriculture and forestry*	*Agricultural machinery research institutes*	*Industrial machinery research institutes*
Province	±	100	6	2	1	2*
Region or prefecture	±	45			13	
County	±	400			88	16†
Communes and brigades	±	4 000 (Groups)			10	

* Machinery Design and Processing Machinery Institutes.

† R and D departments in factories. The number of these departments is expanding rapidly.

SOURCE Data supplied by the various provincial bureaus related to the year 1980 or 1979. Also see Christopher M. Clarke, *China's Provinces: An Organisational and Statistical Guide* (Washington, D.C.: National Council for US–China Trade, 1982), p. 200.

rarely assigned to communes and production brigades. The latter are serviced mainly by the agricultural technicians who are supplied by the county R and D institutes.

We were told (in November 1981) that it was planned to abolish institutions at the prefecture or district level and transfer them to the counties or the province. For example, 11 of the 13 agricultural machinery research institutes are proposed to be transferred to the counties whereas others may be transferred to the province. The objective of such a centralisation of science and technology (which is inconsistent with the policy of decentralisation of economic management examined in Chapter 2) is not quite clear, however. One plausible argument presented to us in Hunan was that of streamlining and avoidance of duplication of effort.

4 Technology Imports and Foreign Investment

The four periods discussed in Chapter 3 can be characterised by the degree of technological isolation or self-reliance (the Great Leap and to a lesser extent, Cultural Revolution periods) or that of dependence on imports of foreign technology (the Soviet and post-Mao periods).

A HISTORICAL PERSPECTIVE

In the 1950s, the imports of capital goods and technical know-how represented an important means of technology transfer to China. During this period, imports accounted for about 40 per cent of total investment. This ratio declined significantly in the early 1960s but recovered somewhat in the 1970s. The import component of investment in recent years (mid-1970s) is estimated at around 10 per cent.[1] However, in the early 1980s this ratio is likely to have increased further although perhaps not as high as the 54 per cent indicated in Table 4.1 below which was based only on 130 industrial project proposals, the final outcome of which was uncertain.

Until the mid-1950s, the Soviet Union was the main and dependable source of supply of foreign equipment and technical assistance. For example, it is reported that during 1953–62, only 68 large construction projects were built by various Eastern European countries compared with 1400 projects built by the Soviet Union or by the Chinese on the basis of Soviet designs and blueprints.[2] During the first Five-Year Plan (1953–7), 156 turnkey projects were introduced in such sectors as energy, metallurgy, chemical industry and defence. With the departure of the Russians in 1960 and in the wake of the Great Leap, the technology imports seemed to have slowed down temporarily. It is reported that the level of technology imports picked up again with the import of turnkey projects from the Western countries during 1964–7. It is estimated that the indices for machinery imports rose from 52 in 1965 (the eve of the Cultural Revolution) to 242 in 1975.[3] Another wave of technology imports was between 1973 and 1977. Chen Weiqin reports that in '1973 the State Council gave permission to introduce from Japan, United States and Western European countries a batch of complete sets of equipment and simple machines'.[4] The original plan was to utilise 4300 million dollars of foreign exchange for this purpose: the plan was thus called the '43' scheme. By 1977, a total amount of 3500 million dollars was spent on foreign contracts. From 1977 to 1979, the share of machinery in total imports rose further: from 15.2 per cent in 1977 to 17 per cent in 1978, and 26.2 per cent in 1979. Of the total machinery imports, non-electrical equipment accounted for the largest share, namely, 34.8 per cent of the total machinery imports in 1977, 44.4 per cent in 1978, and 58.4 per cent in 1979.[5]

During the late 1970s, the rush to import advanced technology for the rapid modernisation of industry necessitated a major drive for the promotion of exports which were required to pay for imports. Thus in 1978, China signed a major trade contract with Japan for the import of plant and machinery in exchange for crude oil and coal. In the same year, another trade agreement was signed with the European Economic Community. The Chinese foreign trade corporations were allowed to sign contracts directly with foreign firms (with an authorisation from the first Ministry of Machine Building) with the aim of rapidly expanding exports. These corporations used tender as a means to obtain technology and equipment from abroad, for example construction equipment, mainly tipper trucks, heavy trucks, jeeps, bulldozers and truck cranes.[6]

The provincial trade corporations cover both manufacturing and trade. For example, the Hunan Provincial Trade Corporation is responsible for such products as machine tools, electrical products, agricultural machinery, instruments and mining and general purpose machinery. The corporation is engaged in cooperative production with foreign firms. In November 1980, it negotiated cooperative production between Electronic Space Systems Corporation (ESSCO) of United States and Xiantan Works for a solar plant. Other examples of Hunan's potential collaboration with foreign companies are:

(i) negotiations with a Belgian firm concerning an underground construction project;
(ii) contacts with Harris Corporation (USA) for cooperation in the production of machine tools; *and*
(iii) negotiations between Bobcock Moxey Ltd (UK) and Changsha General Mining Machinery Plant concerning joint production for bucket wheel reclaimers.

The Hunan Trade Corporation also assisted industrial enterprises in export production according to international quality standards. For example, we were told that the Corporation assisted the Changsha industrial pump factory in selling modern pumps and castings to the Ingersoll–Rand Company of the United States in 1980 (see Appendix II). In July 1979, a State Council decree had provided for such industrial enterprises to participate in foreign trade and retain foreign exchange earned. In a wire and cables factory in Xiantan, the factory manager told us that the central authorities in Beijing had put pressure on provincial industrial enterprises to export rather than to meet the unsatisfied domestic demand for their products. This pressure derived from the need to earn foreign exchange to pay for the importation of machinery from abroad. It was reported that a few enterprises were allowed to keep some portion of foreign exchange earnings on an experimental basis. These earnings were shared between the central government, lower levels of government (provinces and counties) and the export corporation.

In practice, there seemed to have been resistance from the Ministry of Foreign Trade to the State Council directive which in fact diminished the Ministry's role and control over export promotion. Reports indicate that with few exceptions, such as Guangdong and Hunan, the responsibility for foreign trade has been recentralised. Furthermore, very few enterprises share the foreign exchange earned through exports. It is stated that the State Council directives of July 1979 'have not been honoured'[7] due mainly to resistance from the Ministry of Foreign Trade.

Whatever the causes of non-compliance of the new reforms it seems economically more rational to centralise foreign trade and allocation of foreign exchange for different end uses. Decentralisation of responsibility for foreign trade and exports to the provincial and enterprise levels is most likely to lead to a concentration of exports in fields in which China may not necessarily have a comparative advantage. Similarly retention of foreign exchange by individual enterprises prevents its most optimal use for national interests like employment generation through the promotion of labour-intensive products and activities, for example. Furthermore coordination of exports and foreign exchange allocations by a large number of industrial enterprises becomes a very difficult administrative task.

In 1981, the total value of Hunan's exports amounted to 354 million US dollars, that is 8.9 per cent above its annual quota. According to the Director of the Hunan Foreign Trade Bureau, the volume of Hunan's exports increased at the rate of 16 per cent per annum between 1979 and 1981. It is claimed that more than 110 types of machinery and electrical products have been exported. The industrial and mineral products accounted for 63 per cent of Hunan's total export volume in 1981. Agricultural and sideline products were other major export items.[8] In 1981, there were 525 factories and mines engaged in export. Hunan has also set up an export commodity production centre at Xiantan (covering 7 counties, 4.8 million people, and 320 000 ha of farmland) to develop export potential in cereals and oils, light industrial goods, machinery and so on.[9] Special measures were proposed to be undertaken (for example, loans, higher prices for better-quality products, preferential treatment for the import of materials needed for export production and improved transportation services) to improve export performance.

Foreign construction contracts are another means of earning foreign exchange. In 1980, Hunan workers were involved in 15 such construction projects abroad, and engineers from the province worked on 21 design and technical projects in a number of foreign countries. The completion of these projects is expected to yield over 40 million US dollars' worth of foreign exchange.[10]

The Hunan province has resorted to joint ventures, licensing and know-how agreements, and compensation and processing trade to import foreign capital, technology and know-how to modernise its industry. In 1979 and 1980 there were 39 compensation and processing trade contracts, and in 1981, another 19 contracts were signed. The contracts signed in 1979–80, when completed, are expected to earn an income of 4.77 million US dollars. Compensatory trade agreements signed in 1981 alone provided Hunan with 230 000 US dollars' worth of foreign capital and equipment.[11] Hunan traders find compensation trade attractive since they do not have to obtain foreign exchange allocations or loans from the Bank of China. However, it has its possible disadvantages in terms of some risk of unfavourable terms for the Hunan firms in the overpricing of imports, underpricing of exports, or both. In November 1981, we visited a piston rings factory in Changsha which entered into an agreement with a British firm for the supply of equipment and technical know-how. The technical know-how relating to chemical solutions supplied by the foreign company was not available anywhere inside China (see Appendix II).

In the early 1980s, the imports of technology into China have continued. In 1982, 200–300 projects were financed with foreign technology and funds. This is more than twice the annual average figure for the period 1979–81. In 1983, the corresponding figure is expected to rise to 500 projects.[12] The State Economic Commission has recently proposed the import of 3000 technology items over a three-year period to improve productivity in small and medium enterprises. It is claimed that Guangdong factories which used foreign technology components have already improved their export performance. For example, the province exported 33 million garments in 1982, which is twice as many as in 1979. People's Construction Bank in China is expected to extend a loan of 2 billion yuan (about 1 billion US dollars) in 1983 for renovation and modernisation of existing enterprises through technology imports. This represents an increase of 800 million yuan over 1982.[13]

SHARE OF FOREIGN INVESTMENT

By the end of 1981, China had approved 40 joint ventures accounting for 190 million dollars of total investment, with 87.48 million dollars provided by foreign sources. In addition, 590 compensatory contracts have been approved accounting for about 460 million dollars' worth of imported equipment and machinery. It is estimated that the total amount of foreign investment accounted for by the above projects is about 2.9 billion dollars.[14] (The figure is calculated on the basis of amounts agreed instead of those actually used.)

More recently, China made a new bid to seek foreign investment for 130 industrial projects at a China Investment Promotion Meeting which was organised in collaboration with UNIDO (Guangzhou, 7–11 June 1982). Tables 4.1 and 4.2 below indicate the number of projects by the Central–South Provinces and the share of foreign investment required for projects allocated to each of these provinces. Table 4.1 indicates that after Hubei, Hunan's share of foreign investment for its industrial projects (for example, high and low voltage electrical porcelain, metal cut-off bandsaws, and biological microscopes) would be the lowest: it is also much below the share for the Central-South Region or for China as a whole.

Table 4.2 shows the distribution of investment projects by type of industries. As is to be expected, the share of foreign investment is higher in the 'heavy' industries such as metallurgy, chemicals and machine-building industries in which China has to import most of the advanced technology required. Considerable emphasis is also placed on light industry which has been allocated 29 out of 130 projects. The industrial projects in the light consumer goods sector were selected

TABLE 4.1 *Share of foreign investment in industrial projects in the Central–South Region*

Region	*No. of projects*	(1) *Total investment (US$10 000)*	*Average (US$10 000)*	(2) *Foreign investment (US$10 000)*	*Average (US$10 000)*	*Share of foreign investment in total investment (%) (2 ÷ 1)*
Henan	4	1 475	368.7	735	183.5	49.8
Hubei	8	5 965	745.6	2 417	302.1	40.5
Hunan	6	1 887	314.5	861	143.5	45.6
Guangxi	9	34 494	3 832.7	24 519	2 724.3	71.0
Guangdong	2	295	147.5	275	137.5	93.2
Total for the Region	29	44 116	1 521.2	28 807	993.3	65.0
Total for China	130	163 864	1 260.5	89 590	689.2	54.6

NOTE The above data relate to tentative project proposals which were prepared by the new Ministry of Foreign Economic Relations and Trade for the Guangzhou Investment Promotion Meeting (7–11 June 1982). The State Planning and Economic Commissions inspected and approved these proposals before they were presented to the Meeting.

SOURCE Adapted from 'China's New Major Drive to Seek Foreign Investment', *Economic Reporter* (English Monthly) Hong Kong (April, 1982) no. 4.

since they enjoyed a reliable supply of local raw materials and an adequate amount of domestic financial resources. According to Premier Zhao Ziyang, priority is given, in the selection of these projects, to 'small and medium-sized enterprises turning out products much in demand on the domestic and world markets'. It is now an explicit policy of the Chinese government to revamp and expand existing enterprises through quick-yielding foreign investments and imported technology. Only five of the 130 projects therefore relate to newly established enterprises.

At the conclusion of the investment meeting, letters of intent were signed for 70 projects which included 14 for light industry, six for foodstuffs, eight for textiles, three for chemicals, ten for building materials, five for machinery, three for meters and instruments, three for medical apparatus, six for metallurgy, ten for electronics and three for timber processing.

A PRELIMINARY ASSESSMENT OF IMPACT

The debate on self-reliance or technological dependence so common in the Third World today took place in China as far back as the late 1950s when the Russians cancelled their technical assistance and aid agreements. After the departure of the Russian technicians, the Chinese government had no real choice but to follow a path of self-reliance through the utilisation of local resources. With the exception of the Soviet Union, China had no real and willing supplier of advanced technology which meant that she had to resort to a policy of self-reliance. Until then, exclusive dependence on one single supplier of aid and trading partner must have made China vulnerable.

In the 1950s, the speed and direction of technical change in the Soviet Union seems to have largely determined the pattern of technological development in China. While the Soviet aid did certainly benefit China in the process of modernisation, it might have also had the adverse effect of inhibiting indigenous innovations and adaptation of imported technology. One could even argue that the discontinuation of external aid to China was a blessing in disguise in so far as it had a stimulating effect on the development of indigenous technology and the consequential learning effects that accompany such a process.[15] It is debatable whether the local technology systems at the county, commune and brigade levels examined in Chapter 3 would have evolved so much had the imports of advanced technology and technical personnel from the Soviet Union continued.

TABLE 4.2 *Distribution of investment projects by type of industries (China)*

Industry	*No. of projects*	*Total investment (US$10 000)*	*Average (US$10 000)*	*Foreign investment (US$10 000)*	*Average (US$10 000)*	*Share of foreign investment in total investment (%)*
Light industry	29	35 702	1 231.1	19 161	660.7	53.6
Textile industry	11	8 419	765.4	3 842	349.3	45.6
Chemical industry	11	25 235	2 294.1	12 466	1 133.3	49.4
Machine-building industry	19	38 821	2 043.2	26 221	1 380.1	67.5
Building materials industry	21	23 001	1 095.3	10 363	493.5	45.0
Meters and instruments	4	1 690	422.5	610	152.5	36.0
Medical apparatus and instruments	3	660	220	260	86.7	39.3
Metallurgy	11	9 276	843.3	6 267	569.7	67.5
Electronics	18	17 758	896.6	8 620	478.9	48.5
Forestry	3	3 302	1 107	1 780	593.3	53.9

NOTE Data in this table refer to tentative project proposals which were prepared by the new Ministry of Foreign Economic Relations and Trade for the Guangzhou Investment Promotion Meeting (7–11 June 1982) which it organised in collaboration with UNIDO.

SOURCE 'China's New Major Drive to Seek Foreign Investment', *Economic Reporter*, Hong Kong (April 1982) no. 4.

This possible dampening effect of imported technology on indigenous innovations and capacity-building does not seem to have been fully recognised at the beginning of the post-Mao period. A commitment to the 'four modernisations' policy and a drive for rapid export promotion could not be fulfilled without imports of foreign technology in fields in which China did not possess any alternative technology of its own. During 1977–9 therefore, the importation of foreign technology was liberalised. It is now frankly admitted in China that these imports were not only excessive but also sometimes ineffective. To quote a Chinese scholar:

> In the past two years for example, lack of experience has meant that the scale of our imports of complete sets of equipment has been too large and has exceeded our capacity to make good use of them. Furthermore, some of the items imported have not been economically effective.[16]

There are indications from the Chinese press and economic literature that the Chinese planners now recognise that the importation of complete 'turnkey' projects from abroad creates unnecessary dependence and slows down national capacity-building in science and technology. Under the new policy framework, preference is given to 'unpackaging' of technology and to the importation of basic components and materials essential for raising technological levels in China. The most recent policy of import-substitution in some capital-goods sectors is a further indication of some priority given by the Chinese to indigenous capacity-building through protection of its domestic industry. For the assessment of technology imports and foreign investment, the post-Mao period therefore needs to be divided into two sub-periods, namely, the late 1970s (1977–9) and the early 1980s (1980–3). While the former sub-period is characterised by liberal imports of complete 'turnkey' projects and an emphasis on new projects and major expansions, the latter sub-period marks the beginning of unpackaging of technology imports, and a stress on technological transformation in existing enterprises.

It is debatable whether this change is the result of a conscious shift of policy on technology imports. It could well be imposed out of economic necessity. It is quite clear, however, that the massive technology imports, especially in 1978 and 1979, turned out to be extremely expensive. In 1978 alone, contracts worth 6400 million dollars were signed with foreign countries for extremely large projects like the Baoshan iron and steel plant[17] and the Nanjing petrochemical plant. Originally, the planned contracts were estimated to require 12 400 million dollars, but many of these contracts could not be signed due to severe foreign exchange constraints. Moreover the completion and operation of many of these projects further accelerated imports of raw materials (iron ore) and other supplementary materials (chemical and metallic materials). Difficulties encountered in the installation and operation of many imported plants may have further led the Chinese authorities to go slow on technology imports. Chen Weiqin mentions the following major problems:

(i) delays in capital construction (11 of the 24 projects were delayed by one year, whereas one was delayed by three years);
(ii) underutilisation of capacity (several projects operated at below 50 per cent);
(iii) low return on investment (very few projects had 'comparatively good returns' which are defined in terms of recovery of investment in three to four years)[18] and at times even losses (for example, the two chemical fertiliser manufacturing plants in Hubei and Hunan which suffered losses due to an increase in the price of light oils despite full capacity utilisation); *and*
(iv) duplication of equipment and technology imports thus leading to excess capacity in domestic machinery production.

Many of these problems are explained by a lack of a 'serious scientific, comprehensive, balanced and long-term planning' and a 'blind faith in the country's economic and material resources'.

The preceding discussion suggests that the Chinese government's apparent shift in technology policy in the 1980s is an inevitable consequence of a mixed experience with liberal technology imports during 1977–9. But this shift may also have occurred, at least to some extent, as a result of the macroeconomic policies like the shift from capital accumulation to consumption and possible use of foreign exchange earnings for foodgrains imports instead of imports of machinery. The slower rate of investment in the economy in the 1980s (it is planned to be no higher than 25 per cent instead of the actual rate of 30 per cent in 1980) is bound to slow down the growth of technology imports into China. In the case of Hunan, capital investment in 1980 was 31.7 per cent lower than in 1979. For 1981, it was planned to be reduced further by 31 per cent. Work on 78 large projects, each requiring an investment of over 100 000 yuan, was either stopped or slowed down.

This measure is likely to reduce the imports of materials and equipment inputs for these and related projects.[19]

Furthermore, foreign capital is proposed to be utilised more selectively in a limited number of priority areas only. To quote Shang Zijing, Vice-Governor of Hunan:

> The foreign capital at present, will be mainly used to exploit energy resources, develop the production of non-ferrous metals and speed up the technical transformation of the light and textile industries. It will be used to transform the existing enterprises with a view to raising the quality of products and increasing the economic effects.[20]

There is some indication that the most recent technology policy shifts are also guided by the objectives of energy-saving and employment-generation. In our interviews in Hunan in November 1981, most Chinese government officials and scholars emphasised that a shift to small and medium projects, to light industry in general, and a more selective technology importation policy, was intended to economise on scarce capital and energy and to utilise abundant labour. The recent policy of import-substitution may also be motivated, at least partly, by the need to generate additional employment.

5 Agricultural Mechanisation

Rapid farm mechanisation was considered an essential prerequisite for agricultural modernisation in China during the Mao period. A goal was set to implement a policy of fully-fledged mechanisation in 1980. At the first National Agricultural Conference held in October 1975, Hua Guo-feng announced that agricultural mechanisation was to 'be basically achieved by 1980'. The Central Committee of the Chinese Communist party issued a document to reiterate this policy. However, the implementation of this policy did not progress as rapidly as expected. For example, at the Third National Conference on Agricultural Mechanisation held in January 1978, Vice Premier Yu Quili noted that by 1980, only 70 per cent of the major agricultural, forestry, animal husbandry, sideline production and fishery operations should be mechanised, and that this figure should rise to 85 per cent by 1985.[1]

Enthusiasm for rapid farm mechanisation subsided in the post–Mao period. In 1979, the old policy was abandoned in favour of a more selective approach based on local availability of climatic, manpower and financial resources.

A NEW POLICY OF GRADUAL MECHANISATION

The current mechanisation policy is different from the old in both objectives and contents. It is guided, inter alia, by the need to avoid labour displacement and to generate additional employment. The Chinese planners and policy-makers are becoming increasingly conscious of the need to ensure only a gradual farm mechanisation in order to avoid 'a radical displacement of labour because the country's industrial base is not yet able to absorb large numbers of unemployed people'.[2]

In 1975, the American Rural Small-Scale Industry Delegation was struck that the Chinese farm mechanisation policy did not consider its negative impact on the displacement of labour. To quote the Delegation:

> China has a low level of per capita arable land and shares the same basic problems of high population density and subsistence agriculture with other developing countries. Yet, surprisingly, we did not hear a single comment indicating any fears of unemployment through agricultural mechanisation. On the contrary, we consistently found that the Chinese look at mechanisation as an effective tool to improve labour productivity and to release labour for more productive employment.[3]

In 1981, the Chinese position was in contrast to the above statement. According to some estimates, as much as 30 per cent of the rural labour force was in surplus in late 1981.[4] Since a large number of educated urban youth was already waiting to be employed, it would be impossible to reabsorb the technologically displaced rural labour force in the urban sector.

We were told in Hunan that the fear of labour displacement through rapid mechanisation has also influenced the formulation of new policies in the province. For instance, a higher priority is now given to lighter and smaller machines (for example, roto tillers, threshers, and small irrigation equipment), and to manually operated and animal-drawn equipment than to larger and heavier machinery like tractors and combine-harvesters. Small farm tools and implements do not displace labour, are easier to operate and maintain, and are more suited to the hilly terrain of the province. Second, they are also more economical on smaller farms which are growing in the wake of the agricultural responsibility system discussed in Chapter 2.

Table 5.1 shows the degree of mechanisation by the hierarchy and types of Chinese rural institutions. For example, simple farm tools and implements and hand tractors are used mainly at the level of the production teams whereas medium-sized equipment is used mainly at the production brigade level, and large tractors and other farm machinery are generally used at the level of the communes. Recent decollectivisation symbolised by the responsibility system seems to make large tractors and other machinery unsuitable for smaller farms operated by production groups or individual households. Instead, it is likely to stimulate demand for simple equipment which is suited for intensive small-scale farming.

Under the agricultural responsibility system, the smaller size of

TABLE 5.1 *Farm mechanisation by rural institutions*

Institutions	*Responsibility*	*Scale of farm machinery*
Household	Private plots, distribution, consumption among individuals	Small-scale and simple
Production team	Management of agricultural production, ownership of small/medium farms	Small-scale
Production brigade	Primary schools, cooperative health services, small-scale industries	Small or medium-scale
People's commune	Secondary schools, health clinics, small and medium industries, marketing services, collective farms.	Large-scale and heavy machinery

SOURCE Adapted from World Bank, *China: Socialist Economic Development*, vol. I, Washington, DC., p. 55.

farms provides an economic rationale for the new policy of concentration on the local manufacture of small agricultural machinery. In Hunan, we were told that the provincial government was giving top priority to the production of 2.5–5.0 h.p. engines which were hitherto not in production.

The Hunan provincial government accorded a high priority to the gradual mechanisation of rice farming in the first instance. It is proposed to introduce mechanical equipment for irrigation, drainage, forestry and so on only at a later stage. This selective and gradual approach may be guided by the lack of adequate trained manpower. The government recognised that training of a large number of technicians, tractor drivers and operators would be essential before agricultural mechanisation could be accelerated. Training of new operators quickly is complicated by the lack of specialised production which is manifested in the existence of a large variety of equipment models.

IMPLICATIONS OF NEW POLICY

The new farm mechanisation policy is likely to lead to two important consequences. In the first place, it is creating excess capacity in the agricultural machinery industry which it is proposed to convert into the production of light consumer goods.[5] Second, a higher priority to simple farm tools and implements should induce an expansion of rural industry at the brigade and team levels. However, neither of these two consequences will follow without problems. First, switching of capital capacity from one industrial use to another depends partly on the degree of the multiple-purpose nature of the equipment. All heavy equipment capacity is unlikely to be converted into the production of light consumer goods without sacrificing efficiency. Neither is it easy to switch it to the manufacture of simple tools and implements. Thus there are press reports about continued shortages of simple equipment like light-weight power tillers.[6] Although the new policy seems to induce the expansion of such light machinery as threshers in the rural areas, this machinery is designed by farmers and mechanics with practical experience but no engineering expertise in machinery design and development. The result is the production of equipment of very poor quality.

Considering the above limitations, it is unlikely that China will be able to design and manufacture small agricultural machinery quickly. In the absence of an established base for small machinery manufacture, it may be some time before China becomes self-sufficient. The Chinese policy-makers seem quite aware of these difficulties. This is clear from the following news statement made by Yang Ligong, Chinese Minister of Agricultural Machinery:

> It will take some time to design and produce in batches some small agricultural machinery that suits small-scale farming and diversified undertakings. It will also take some time for the peasants to save enough money to buy agricultural machinery. The readjustment of the agricultural machinery industry and the reform of the supply and marketing of the agricultural machinery has just begun. In this situation, declining agricultural machinery sales are inevitable, but this only affects some regions and some agricultural machinery.[7]

The reforms mentioned by the Minister in the above statement include: (i) identification of priority products which will receive major attention in future; (ii) determination of a proper balance between quantity and quality of production; and (iii) establishment of a general company for the control of supply, marketing and servicing of farm machinery. (The company also helps the enterprises to organise customer interviews and farm machinery exhibits in addition to assisting in the signing of sales contracts.) These reform measures are essential to solve, inter alia, the acute problem of low-quality production, underutilised equipment, and poor maintenance. In many cases, no more than 50–60 per cent of the farm machinery is in good condition. The operating capacity of a good many tractors is as low as 100 standard *mou* (or 6.7 hectares) per horsepower. Furthermore the production and maintenance costs per mechanised *mou* vary from 0.3/0.4 yuan to over 1 yuan.[8]

It is rather peculiar that the number of the farm-level stock of tractors has recently increased in China despite the new policy of gradual and selective farm mechanisation. This phenomenon may be explained by a number of plausible factors. First, it may reflect investment in transportation. The use of tractors for short haulage (0.5 to 1.0 ton capacity trailers) accounts for 80–90 per cent of the total tractor operating time. Second, the ploughed area per tractor has recently declined from 190 hectares in 1965 to about 60 hectares in 1976 and only 34 hectares in 1980. Third, trucks suitable for road transport of agricultural materials are expensive for purchase by the communes. Fourth, fuel pricing policy has also favoured tractor transportation: diesel fuel used by tractors is available at special subsidised rates for agricultural purposes (these rates have not been changed since 1965) compared with gasoline used by trucks which is sold at higher prices. Trucks are also at a disadvantage since they are subject to heavy annual licensing fees.[9]

EXTENT OF MECHANISATION

Some indication of the degree of farm mechanisation can be given by the farm area cultivated by machinery and by the output of farm machinery industry.

Taking the former indicator first, the area under tractor cultivation is 40 per cent of the total farmland area in China. Machinery is used in nearly 55 per cent of the irrigated farm lands. The processing of agricultural and sideline products has also been mechanised considerably.[10]

In the case of Hunan, 25.4 per cent of the area under cultivation in 1980 was harrowed by machines. This implies that three-quarters of the cultivable area continued to be ploughed and harrowed by non-mechanical methods. The area covered by mechanical threshing of early and middle-season rice was a little higher, that is, 37.5 per cent of the total acreage. However, the fact that the processing of foodgrains and industrial crops (for example, cotton, oilseeds and tea) is gradually being mechanised suggests a potential growth in the demand for small and simple tools and implements within the rural areas. This growing demand should induce rural manufacture of farm implements. The Hunan experience indicates that rural industrialisation tends to promote the demand for farm equipment. Therefore farm mechanisation is spread more rapidly in those regions where rural industry is more developed than in others where it is still in its infancy. Commune rural industry is able to absorb farm labour released by mechanisation. It is also an important source of capital for investment in farm machinery in the absence of any significant state investment. On the supply side, the rural industry provides capacity for manufacture, repair and servicing of farm machinery. This interaction between farm mechanisation and rural industry is illustrated in Figure 5.1, which also indicates that farm mechanisation has been necessitated by the need to reduce human drudgery, raise labour efficiency and provide against natural disasters.

The degree of mechanisation in agriculture in the post-Mao period needs to be examined in the light of the emergence of a very different rural institutional structure. One also needs to take into account the different situation regarding labour utilisation. Even during the Mao period, farm mechanisation has been more rapid in the North-East region of China which is characterised by the large size of plots and sparse population compared to the conditions in the Southern, more populated region with smaller holdings. Nevertheless rapid mechanisation was also promoted in the Central–South Region, presumably to release labour for non-farm activities and to improve working conditions. The situation, particularly since 1979 with the introduction of the agricultural responsibility system, is likely to have changed.

In 1979, the Chinese Farm Machinery Industry produced over 667 000 large and medium-sized tractors, 1.6 million hand tractors, and 2 912 000 grain-processing machines (see Appendix I, Table A.5). The stock of these major farm machinery products is reported to have increased by more than 10 per cent over 1978.[11] In the case of Hunan, the number of large and medium-sized tractors has increased from 28 in the initial post-liberation period to over 18 000 in 1979. In addition, there are about 56 000 hand tractors and 147 000 grain-processing machines (see Appendix I, Table A.5). It is also reported that at present about 100 000 manual rice threshers, 700 000 or more hand sprayers and 50 000–60 000 small diesel engines are produced annually in Hunan.[12]

Farm mechanisation in Hunan does not seem to have advanced as

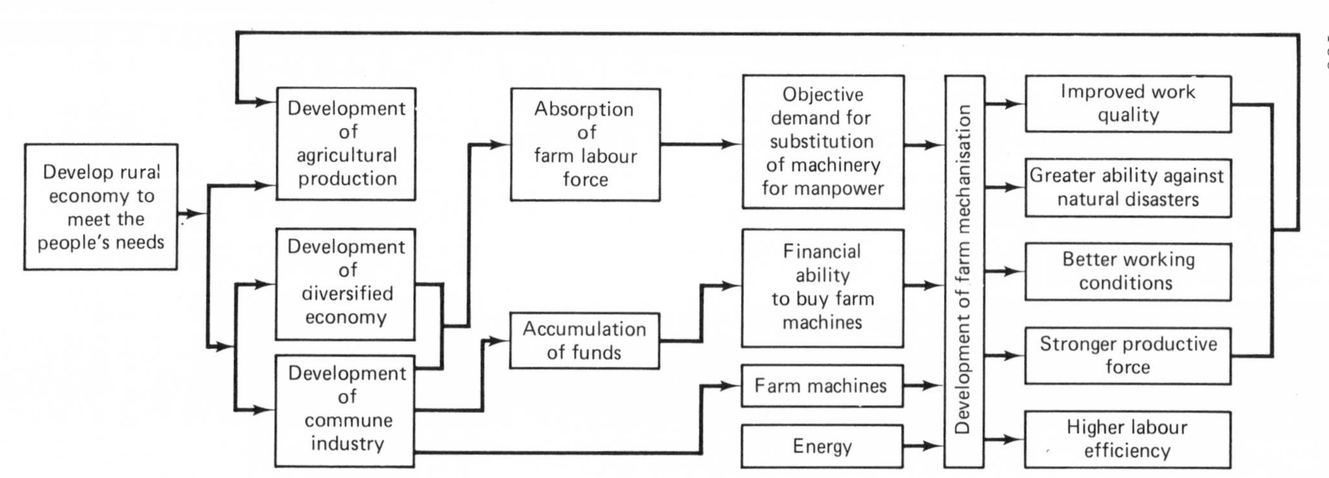

FIGURE 5.1 Interaction between farm mechanisation and rural commune industry.

SOURCE Zhu Lunkun *et al.*, *Interaction Between Farm Mechanisation and Commune-run Industry in Shanghai Rural Areas*; paper prepared for the Seminar on Mechanisation of Small-Scale Farming held in Hangzhou, People's Republic of China, 22–26 June 1982.

much as in other provinces of the Central–South Region. In order to examine the relative importance of farm mechanisation in these provinces, data were obtained relating to the total supplies of different types of agricultural machines in each of these provinces as well as in China as a whole. Table 5.2 estimates the number of large tractors and hand tractors per hectare of arable area, irrigation machinery per hectare of irrigated area, and grain-processing machinery per unit of grain output in the five provinces. It also shows the man–land ratio in each province, the agricultural output value per capita, and the share of the stock of each type of machinery for each province in the total stock for China as a whole.[13]

The ratios of tractors per hectare show that tractorisation in Hunan is below the average for the Central–South Region: it is also less marked than that in Guangdong, Guangxi and Hubei. This is true only partly for mechanisation of irrigation where Hunan seems more advanced than the above three provinces. Mechanisation of grain-processing is at present not very advanced in Hunan. Except Guangdong province, Hunan is ranked lowest in the ratio of grain-processing machinery to output.

The second indicator of farm mechanisation, that is, the ratios of farm machinery, further reinforces the above findings. The share of Hunan in the total production of farm machinery in China is quite low particularly for tractors, compared with the shares of other provinces of the Central–South Region. The ratios for Hunan are also below the average ratios for China as a whole, with the exception only of hand tractors, the ratios for which are very similar for Hunan and China.

To some extent the suitability of different types of farm machinery also depends on the type of terrain. It is quite possible that some of the variations in the ratios presented in Table 5.2 are explained by this factor. For example, the low share of tractors in Hunan may be explained by its mountainous terrain.

The ratios of agricultural machinery per hectare of land also vary a great deal among the Asian countries. This is illustrated in Table 5.3 below which compares Chinese data with that of India, the Republic of Korea, the Philippines and Thailand.

The ratio of tractors per 1000 ha for China is higher than that for other Asian countries like the Philippines and Thailand. Similarly, the ratio of power tillers is also higher. But the Chinese ratio for tractors and power tillers together is lower than that of the Republic of Korea. Comparison between the two sets of ratios shows greater reliance on power tillers which implies relatively greater emphasis on smaller-scale

TABLE 5.2 *Relative importance of agricultural mechanisation: Central–South Region (1979)*

Province	*Man–land ratio**	*Cropping index (%)*†	*Agricultural output value per capita*‡	*Agricultural machinery:* *Large and medium tractors*	*Hand tractors*	*Irrigation machinery*	*Grain-processing machinery*
				(per 000 ha arable area)	*(per 000 ha arable area)*	*(h.p. per 000 ha irrigated area)*	*(units per 000 tons of grain output)*
Henan	9.23	152.9	143	7.3	14.8	2 150	14.4
Hubei	11.15	207.1	203	8.8	26.1	1 150	11.1
Hunan	14.11	242.3	180	5.2	16.3	1 227	6.6
Guangxi	12.48	194.2	138	8.4	33.9	610	8.6
Guangdong	15.52	217.4	143	6.2	37.0	1 096	4.5
Central–South Region	11.84	–		7.2	23.2	1 392	9.2
China	8.47	149.2	163	6.7	16.8	1 982	8.7
Share of each province in the total stock of Machinery in China as a whole (%)							
Henan				7.8	6.3	11.0	10.5
Hubei				4.9	5.9	3.8	7.1
Hunan				2.7	3.4	4.2	5.0
Guangxi				3.3	5.3	1.3	3.5
Guangdong				3.0	7.1	3.3	2.7
Central–South Region				21.7	28.0	23.6	28.8
China				100.0	100.0	100.0	100.0

* Derived from Table 1.2, column 3 and Appendix I, Table A.4, column 1.
† Taken from Appendix I, Table A.4.
‡ Taken from Table 1.2, column 4.

TABLE 5.3 *Tractors and power tillers in selected Asian countries*

Country	*Year*	*Cultivated area (000 ha)*	*Tractors*		*Power tillers*		*Total*	
			Total no.	*Units per 000 ha*	*Total no.*	*Units 000 ha*	*Total no.*	*Units per 000 ha*
China	1978	105 000	557 000	5.3	1 370 000	13.0	1 927 000	18.3
	1979*	99 490	667 000	6.7	–		–	
India	1975	161 940	–		–		227 668	1.4
Republic of Korea	1975	2 300	–		–		25 000	27.3
Philippines	1976	7 934	18 049	2.3	32 222	4.1	50 271	6.3
Thailand	1976	11 267	16 000	1.4	90 497	8.0	106 497	9.4

* Taken from Table 5.2 and Appendix I, Tables A.4 and A.5.

SOURCE Shigeru Ishikawa, *Essays on Technology, Employment and Institutions in Economic Development*, Comparative Asian Experience, Economic Research Series no. 19 (Tokyo: Institute of Economic Research, Hitotsubashi University, 1981) p. 149.

equipment. The cross-country differences seem to be due to a number of factors, viz. scale of farming, cropping patterns, agrarian systems and economic incentives. These factors can work in different degrees and directions. For example, a composite index of the degree of mechanisation (that is, tractors plus power tillers) is the highest for the Republic of Korea followed by China and Thailand. Leaving aside Korea, in the case of China, the greater use of tractors and power tillers despite high labour-intensity could be explained by the economies of scale due to collective organisation of agriculture. However, the role of scale economies is likely to be reduced with a virtual elimination of communes.[14]

Man–land ratio could also provide some indication of the degree of farm mechanisation. Other things being equal, one would expect a high man–land ratio to be associated with a low degree of mechanisation. In Table 5.2, the highest man–land ratios in Guangdong and Hunan respectively show the lowest ratios of large and medium tractors per hectare. However, the tractor ratio for Hunan is lower than that of Guangdong although it should be higher. Different cropping patterns and cropping intensities may distort the expected negative relationship between man–land ratio and tractorisation. The diversity of natural and economic conditions in China accounts for differences in cropping patterns and multiple-cropping systems. This is partly illustrated by the wide range of cropping indices for the five provinces in the Central–South China. Hunan and Guangdong have the highest indices respectively. These high indices seem to explain their high man–land ratios and low ratios for tractorisation.

One of the main objectives of farm mechanisation in China is said to be the increase in agricultural production to meet the growing food demand. It is therefore useful to compare the agricultural output per capita with indicators of mechanisation given in Table 5.2. The relationship between output value and tractor density is not as clearcut as one might expect in a market economy. For example, in the cases of Guangdong and Hunan, tractor ratios are lower than in Guangxi although the output values per capita are higher. Also the ratios of hand tractors per hectare are the lowest in Hunan after Henan, yet its agricultural output per capita is the second highest in the region after Hubei. Thus tractorisation in Hunan is quite low.

The reasons for this low degree of agricultural mechanisation in Hunan may be several. First, Hunan is mainly a rice-growing province. Large and medium-sized tractors may not be very suitable for paddy fields; their mobility is particularly limited when the fields are wet. Second, the price of tractors seems to be high considering the high costs of production, low quality, and the high running and maintenance costs due to poor training of farmers in operating the equipment. The economic conditions of the farmers, given the low per capita incomes, may limit their demand particularly for large and medium-sized farm machinery.

MECHANISATION AND EMPLOYMENT

Farm mechanisation can also have positive as well as negative impact on the direct and indirect labour inputs in agriculture. It is not clear whether the Chinese government, prior to 1978, explicitly considered this factor before promoting mechanisation. Agricultural mechanisation was introduced mainly to facilitate double- and multiple-cropping of rice. Ishikawa[15] has compared the Chinese experience during the 1950s (1956–8) with double-cropping of rice without mechanisation to that in the mid-1960s when agricultural mechanisation was accelerated. In the 1950s, double-cropping of rice led to an increase in labour input as well as an acute labour shortage during agricultural peaks. In the absence of mechanisation to reduce this seasonal labour demand, the per hectare yields of the second crop were reported to be very low. However, in the mid-1960s the area under double-cropping expanded considerably in Hunan and other provinces in the Yangtse River Valley (viz. Anhui, Zhejiang, Hubei, and Jiangsu). One of the main difficulties in the double-cropping in this valley is the short rice-growing season (of about 120 days) which makes the agricultural operations such as the harvesting and threshing of the first crop overlap with tillage and transplanting of the second crop. Thus mechanisation of some of these operations by the use of tractors, power tillers and threshers alone could have made double-cropping of rice economically feasible by reducing excessive labour demand at peaks.

It follows that agricultural mechanisation can lead to labour-using and land-augmenting technical change, especially if it is not carried beyond the point where it lowers the amount of annual per hectare labour input for total agricultural production.

To test Ishikawa's hypothesis, discussed above, one would need to know labour input requirements by crops during the mid-1950s and mid-1960s. Some estimates show that labour requirements per hectare have increased between 1929–30 and 1958 as follows: from 195 to 525 mandays for rice, 65 to 210 mandays for wheat, 58 to 253 mandays for corn and 132 to 271 mandays for cotton.[16] More recently, the American Wheat Studies Delegation which visited China in 1975 reported that labour input per sown hectare of wheat, rice and other crops ranged between 225 to 450 mandays, or an average of over 300 mandays.[17] This shows a considerable increase over the 210 mandays reported for 1958. Estimates of labour requirements by crops in Hunan for 1979 are given in Table 5.4 which shows that labour inputs for early and late rice are much higher than the estimates of the American Delegation cited above. These requirements also appear to be a little higher than those estimated by Wiens for Suzhou Prefecture in Jiangsu Province (495 mandays for rapeseed and 509 mandays for single-crop rice).[18] The cotton crop requires a phenomenally high labour input in Hunan, that is, 1105 mandays per hectare compared to 271 mandays reported by Dawson. Rawski argues that the labour input data by crops estimated by the American Delegation do not take account of any increase in the index of multiple-cropping.[19] However, it would appear that data for Hunan do allow for multiple-cropping without which much higher estimates than those of the American Delegation are difficult to explain.

The labour requirements per hectare in cash crop (cotton) are nearly twice as high as those for grains (late and early rice) in the case of Hunan. This suggests that a switch from the planting of grains to cash crops should have a positive employment effect. In fact, the new agricultural pricing policy offers an incentive to producers to shift from grain to cash crops like cotton and oilseeds. In March 1979, the cotton procurement price was increased by 15 per cent: it was increased by an additional 10 per cent in 1980 since the production of cotton did not show any significant increase in 1979.[20]

While the labour requirements for cotton cash crops are higher, the input costs, both material and labour,[21] are also higher than those for early and late rice (see Table 5.5). Under these circumstances, an agricultural pricing policy may be a necessary but not a sufficient condition to induce the producers to shift to cotton. For instance, it is reported that 'past efforts to stimulate cash crop production failed when the government did not deliver needed food supplies'.[22] Thus a package of incentives including favourable prices, guarantee of foodgrains supplies, fertiliser subsidies and so on would be necessary to stimulate cash crop production and generate additional farm employment.

Multiple-cropping is reported to have raised labour requirements

TABLE 5.4 *Labour requirements by crops in Hunan (1979) (mandays per hectare)*

Type of activity	*Crop*			
	Early rice	*Late rice*	*Cotton*	*Rape seed*
Total labour	687.55	541.5	1 105.5	372.75
Direct labour	574.81	446.1	867.0	314.1
Land preparation	180.36	103.65	94.5	69.7
Seed preparation and nursery	90.3	88.35	82.5	45.9
Weeding	11.1	7.95	211.5	25.35
Fertiliser application	102.7	59.7	100.5	59.25
Drainage irrigation	23.08	21.3	7.5	5.1
Management	60.57	49.95	–	–
Plant protection	17.25	24.15	124.5	11.4
Harvesting	89.35	91.05	240.6	62.55
Indirect labour	112.74	95.4	238.5	58.65
Compost accumulation	44.07	31.2	81.0	24.3
Supervision	17.39	19.05	18.0	4.05
Agricultural land construction	32.55	27.75	37.5	26.4
Other	18.6	17.4	102.0	3.9

NOTE Mandays per *mou* have been converted into mandays per hectare.

SOURCE Data supplied by the Hunan Provincial Bureau of Agriculture.

by 60 to 70 per cent in rice-growing areas.[23] Ishikawa has argued that 'the number of labour days applied per year per hectare of cultivated area has also been increased by mechanisation'.[24] For example, in Tachai brigade, 300 days were reported for a single crop of corn, 'which seems to have doubled after two crops of corn and wheat were introduced'. In 1974, the number increased to a phenomenal level of 1002 days including labour input for land improvement.

In the case of Hunan province, the multiple-cropping index increased from 141 in 1952 to 242 in 1979 (see Table 5.6 below). It is noted that this index is much higher for Southern and Eastern China than that in Central (where the index ranges between 160 and 200) and Northern (where the index on an average is 119) China.[25] The much higher index in Southern China including Hunan may be due to the use of plant varieties that mature in a shorter time and are resistant to more extreme weather conditions than those prevailing in other regions. For instance, in Hunan the sown area under green manure grew as a sizeable proportion of the total sown area. Thus the multiple-cropping index based on sown area excluding green manure is much lower than that calculated on the basis of total sown area including foodgrains and green manure.

The much higher cropping index for Southern China including Hunan is likely to be accompanied by a much higher labour input due partly to the preponderance of rice-growing in the region. As noted earlier, cotton and rice require a much higher labour input per hectare than any other crop. While figures are not available, the labour requirement for green manure may also be rather high.

Recent debates in the Chinese press suggest that multiple-cropping in many areas has been overemphasised, pushing it to the point of negative marginal product.[26] However, it is difficult to assess clearly the extent to which the marginal product has declined due to an over-emphasis on multiple-cropping or on a shift from double- to triple-cropping. Some impressions can be gained from a comparison of the change in the cropping index with that in agricultural output per capita, and the cost per unit of output. Data on these variables are not available over time for Hunan or other Southern provinces. However, a recent survey in Suzhou Prefecture shows that material plus labour costs increased from 0.238 yuan per kg in 1966 with double-cropping to 0.248 in 1978 with triple-cropping. The costs per net processed grain are still higher if corrections are made for higher processing losses under triple-cropping.[27] The grain produced per labour day declined from 7.1 kg under double-cropping in 1966 to 6 kg under triple-cropping in 1978. The labour requirements under double-cropping were 1086 days per hectare compared with 1630 days per hectare under triple-cropping.

It is reported that during the Mao period, intensive cropping systems were sometimes forced upon collectives by administrative fiat. The introduction of the household responsibility system in the post-Mao period is likely to do away with these earlier practices. Since production units can now make their own management decisions, and households and even individuals can undertake farming of their own choice, the scope for inappropriate cropping systems would be much more limited than in the past.

In general, in China excessive labour application on a limited area of land is accompanied by diminishing returns to labour. The increase in grain production has not always led to an increase in peasant incomes.[28] This is explained by the significant increases in production

TABLE 5.5 *Input costs by crops in Hunan Province (1979) (yuan per hectare)*

Type of input	*Crop*			
	Early rice	*Late rice*	*Cotton*	*Rape seed*
No. of production teams investigated	5	5	4	5
Land area investigated (ha)	38.08	43.23	22.30	9.37
Production (kg/ha)	5 220	4 567.5	690	997.5
Total material costs (yuan/ha)	622.65	613.65	741.15	263.55
Of which:				
Seeds	46.50	45.00	27.30	2.25
Fertiliser: *of which*	347.85	320.40	418.50	189.30
Chemical fertiliser	166.95	208.35	184.80	6.90
Agro-chemicals	19.65	56.70	122.25	8.55
Draught animals	37.20	27.00	3.00	16.50
Machine ploughing	5.85	6.45	2.55	–
Drainage irrigation	7.05	19.20	23.55	–
Small tools purchase	11.10	7.05	7.65	3.30
Cost of repairs	9.15	10.95	7.20	8.10
Additional cost for initial processing (preliminary)	–	–	42.15	18.45
Other direct costs				
Depreciation	29.25	30.45	31.95	2.10
Management	3.75	3.45	8.10	–
Agricultural land construction	30.75	22.20	–	–
Miscellaneous	1.80	3.30	5.40	11.25
Other indirect costs	72.45	61.50	26.55	3.75
Labour costs	440.25	346.50	873.30	213.90
Total costs (materials plus labour)	1 062.90	960.15	1 614.45	477.45
Cost per 100 kg	20.36	21.02	233.97	47.86
Sale price per 100 kg	25.78	25.78	300.04	74.30
Implied labour cost per manday (yuan)*	0.64	0.64	0.79	0.57

* Figures derived by dividing labour costs by the number of mandays required which are given in Table 5.4.

SOURCE Data supplied by the Hunan Provincial Bureau of Agriculture.

TABLE 5.6 *Hunan's multiple-cropping index over time (1952–79)*

Year	*Cultivated area (10 000 mou)*	*Sown area (10 000 mou)*	*Of which:* Foodgrains	*Of which:* Green manure	*Sown area excluding green manure*	*Multiple cropping index (%)*
1952	5 518	7 804	6 331	691	7 113	141.4* 128.9†
1957	5 802	10 262	8 375	858	9 404	176.8* 162.1†
1965	5 416	10 965	8 318	1 530	9 435	202.5* 174.2†
1970	5 274	11 452	8 077	2 610	8 842	217.1* 167.6†
1975	5 202	12 339	8 568	2 790	9 549	237.2* 183.6†
1979	5 161	12 502	8 556	2 630	9 872	242.2* 191.3†

* Multiple index including green manure.
† Multiple index excluding green manure.

SOURCE Data supplied by the Hunan Provincial Bureau of Agriculture.

costs accompanied by stabilised prices of crop products until the late 1970s. Farm prices were raised in 1979 which seems to have raised peasant incomes since then. It is also reported that uneconomic multiple-cropping is now being discouraged.

To conclude, the positive effect of farm mechanisation on increases in labour input for total agricultural production would be difficult to determine, since multiple-cropping and labour utilisation are caused by a number of factors, namely, farm mechanisation, and biological innovations like chemical inputs and early maturing plant varieties. A proper isolation of the effects of these factors is not possible with the available data.

6 Rural Industrialisation

The development of Hunan's rural industry is intended to supplement agriculture and facilitate its gradual mechanisation. While agricultural mechanisation provides an important link between farm and non-farm activities, repair and maintenance of farm machinery and sub-contracting link rural industry to large urban industry. It is therefore logical to examine, in this chapter, the problems of and the growth prospects for Hunan's rural industries.

CONCEPTS AND DEFINITIONS

The Chinese rural industry is generally defined as covering local rural industrial enterprises that are operated by the collectives at the county, commune, brigade and household levels. These enterprises may be collectively owned or operated/controlled jointly by communes and brigades, or more recently (since the new economic reforms introduced in 1979), by communes and individual members as share-holders in joint stock arrangements.

The conceptual and definitional questions regarding 'rural' industry have bedevilled scholars in non-socialist countries. Nor is the socialist economy of China free from conceptual problems in defining the boundary line between 'rural' industry on the one hand and 'light' and 'urban' industry on the other.

The term rural industry covers both 'modern' rural enterprises using equipment and power and traditional handicrafts 'combining labour with few simple tools and implements without the aid of mechanical power'.[1] Rural industry can also be both light and heavy.

There seems to be no *official* definition of rural industry. But the information available recently from China suggests that it covers rural enterprises operated by communes and brigades (*shedui qiye*) as well as those operated by the Bureaus of Commune and Brigade Enterprises of the District (*qu*) and County (*xian*) governments, or the street industries in small local towns (*zhen*) supervised by the same Bureaus.[2]

However, the importance of these latter enterprises and their weight in the rural commune and brigade industries is rather limited.[3]

The collective (commune and brigade) enterprises are very different from the state industrial enterprises in terms of ownership, the level of technology and the size of plant. For example, the former enterprises require lower capital investment and a shorter gestation period, and lower labour costs and administrative overheads. However, the two types of enterprise may be very similar in nature in the light of other criteria, viz. the kind of products that support agricultural production, the characteristics of the surrounding product and factor markets, the desirability of appropriate technology using local materials and skills, and the importance of the employment and foodgrains problems to be considered in their establishment.[4]

One classification of rural industries is based on the types of product, location of consumption and economic sector. Thus the rural industries are grouped as follows:

(i) those which serve agricultural production, for example, farm machinery, fertilisers and chemicals;
(ii) those which serve industrial production, for example, processing of industrial raw materials, machine spare parts, mining and chemical materials for large urban industry;
(iii) consumer goods industries for daily consumption, for example, food-processing, paper-making, construction materials and canned foods;
(iv) export industries, for example, firecrackers, embroidery, handicrafts and fine arts.

The requirements of these different groups of rural industries for inputs, and financial, material and institutional support from the collective economy or the state administration, vary a great deal. For example, food-processing industries for local consumption are, in general, likely to be based largely on the use of local resources thus requiring a minimum of input from the state administration.

Another grouping of industries done by the Hunan Bureau of Commune and Brigade Enterprises consists of the following categories:

(i) *light industries*: processing of agricultural products, art work, textiles, clothing, leather, paper-making, glassware, and so on.
(ii) *heavy industries*: metallurgy, mining, electrical power, machine building;

(iii) *chemical industries*: chemicals, fertilisers, insecticides, chemical goods for daily use;
(iv) *building materials industry*: cement and cement products, limestone, tiles;
(v) *others*: forestry, electronics.

This last grouping overlaps considerably with the earlier ones but is described more in terms of the number and administrative structure of the provincial bureaus that are responsible for different groups of industries.

While the light and heavy industries are distinguished according to the nature of the product (that is, whether it is a consumer good, an intermediate product or a capital good), the 'rural' industry is usually defined in terms of the levels of administrative control (county, commune, brigade or household) regardless of the nature of the product.[5] A number of heavy industries are also located in rural areas in Hunan. These industries are operated by the communes and brigades. Thus they are partly the responsibility of the Bureau of Commune and Brigade Enterprises, and partly that of the relevant *industrial* bureau in the province.

Second, while 'rural' industry is operated entirely collectively (by communes or brigades), light industry plants may be in both the collective and the state sectors.

In order to get a clearer idea about the overlaps between different classifications, it is useful to examine how the Chinese authorities define light industries.

Light industries cover mainly consumer goods production. It appears from the list of industries labelled as 'light' that they are generally low in capital-intensity and operate at a small scale of production. They are classified into the three following categories:

(i) *first-category light industries*: these industries are related to the fulfilment of 'essential' daily needs of the population, for instance, salt, edible oils, cotton yarn and textile fabric. The purchase, sale and procurement of these goods is controlled by the central government;
(ii) *second-category light industries*: these industries generally supply wider domestic and export markets than the first-category industries. Examples of these industries are bicycles, radios, sewing machines, and so on;
(iii) *third-category light industries*: these are all the industries other

than those mentioned above which are controlled at the city or *xian* (county) level.

The overlaps between rural industry and different categories of light industry are illustrated in Figure 6.1 which also distinguishes between the state, collective and private sectors.

The foregoing discussion of the different classifications of 'rural' and 'light' industry in China suggests some ambiguity in her approach to rural industrialisation. On the one hand, it appears that the policy towards rural industrialisation advocates transplantation of 'light' and 'medium' industry from urban to rural areas. There are indications that many such industries located in rural areas have stronger links with the urban areas than with agricultural and/or non-farm activities within the rural sector. We came across a number of commune- and brigade-operated rural enterprises (such as bicycle-making and pottery) which were linked to the urban industrial sector through the subcontracting arrangements and/or product markets. They had little direct link with the ongoing agricultural activities in the commune or the collective farm in which they were located. In the absence of backward linkages, this approach to rural industrialisation can still lead to an increase of incomes and employment for the rural population in communes and brigades.

It appears that the above 'locational' approach to rural industrialisation in China is secondary to the objective of promoting rural industries with strong links to the rural sector (for example, farm machinery, agricultural processing, repair and maintenance and rural transport).

SOME ECONOMIC FACTS

Detailed information on Hunan's rural industries is hard to come by. The Provincial Bureau of Commune and Brigade Enterprises does not collect systematic information on types of enterprises by level of administration, technology, skilled manpower and employment, raw materials and power requirements, and output and value added. It is therefore impossible to provide a detailed economic analysis of Hunan's rural industrial structure. However, some information is available on the number of commune and brigade enterprises, and on total employment and output of rural industry in China, Hunan and other provinces of the Central–South Region. In the light of this data, some impressions can be gained on the importance and growth

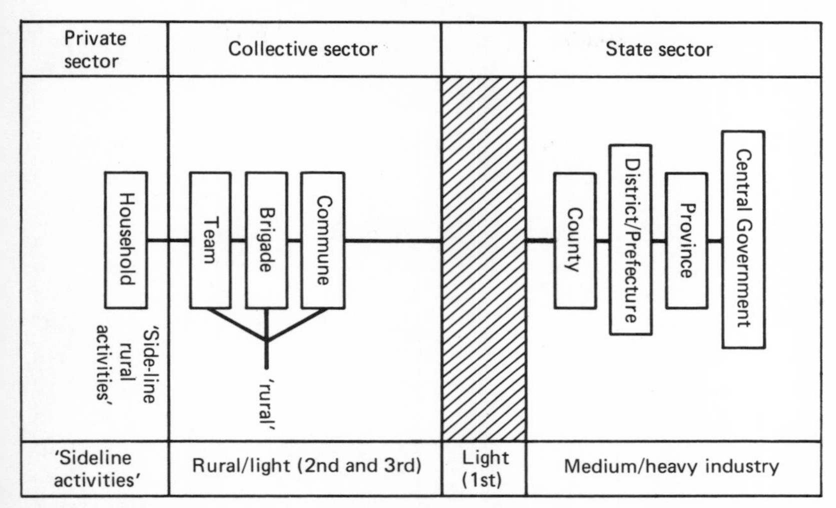

FIGURE 6.1 *Rural, 'light' and 'heavy' industry by levels of administration*

NOTE The shaded area indicates state-owned light industry of the first type which may be the responsibility of the province, district and sometimes also county.

potential of commune and brigade industrial enterprises. These data are analysed below.

Employment and Output Growth

The number of commune and brigade enterprises (including industrial enterprises) has increased considerably in China since the early 1970s. It is reported that by the end of 1979, 98 per cent of the rural communes and 82 per cent of the production brigades had established these enterprises. In absolute terms, there are 1430 million enterprises in China, of which 320 000 were operated by the communes and 1 160 000 by the production brigades (see Table 6.1).[6] Since there are 53 000 communes and 699 000 brigades (see Appendix I, Table A.6),[7] on average there are six enterprises per commune and 1.7 enterprises per brigade. Industrial enterprises, in the case of both communes and brigades, represented the major proportion, accounting for 58.4 and 50 per cent respectively. As is shown in Table 6.1, their shares in total output and employment were also the most dominant.

In 1977, there were 23 million workers employed by these enterprises, contributing a gross output value of 39 billion yuan.[8] In 1979, total employment in these enterprises rose to 29 million with an output contribution of over 49 billion yuan. (In 1980, the total value of gross

TABLE 6.1 *Number, employment and output in commune and brigade enterprises by sector (all China) (1979)*

Sector	*Commune enterprises*			*Brigade enterprises*			*Total (commune plus brigade) enterprises*		
	Number (000) %	*Employment (000) %*	*Output (100 million yuan)*	*Number (000) %*	*Employment (000) %*	*Output (100 million yuan)*	*Number (000) %*	*Employment (000) %*	*Output (100 million yuan)*
Agriculture	59 (18.5)	1388 (10.5)	10.7 (4)	385 (33.2)	3942 (24.7)	27.8 (12.6)	444 (30)	5330 (18.2)	38.5 (7.8)
Industry	187 (58.4)	8416 (64.1)	208.9 (77.4)	580 (50)	9727 (61)	163.2 (73.8)	767 (51.8)	18144 (62.5)	372.2 (75.8)
Transport	26 (8.1)	661 (5.1)	13.7 (5.1)	56 (4.8)	508 (3.2)	9.3 (4.2)	82 (5.6)	1169 (4)	23.0 (4.7)
Construction	23 (7.2)	2253 (17.1)	27.0 (10)	27 (2.3)	731 (4.6)	8.0 (3.6)	49 (3.3)	2984 (10.3)	35.0 (7.1)
Others	25 (7.8)	426 (3.2)	9.6 (3.5)	112 (9.7)	1041 (6.5)	12.9 (5.8)	138 (9.3)	1466 (5)	22.4 (4.6)
Total	320 (100)	13144 (100)	269.9 (100)	1160 (100)	15949 (100)	221.2 (100)	1480 (100)	29093 (100)	491.1 (100)

NOTE Figures in parentheses indicate percentages.

SOURCE *China Agricultural Yearbook* (*Zhongguo Nongye Nianjian*) 1980.

TABLE 6.2 *Changes in number, employment and output in commune and brigade enterprises by sector (all China) (1978–9)*

Sector	*Percentage change*		
	Number	*Employment*	*Output*
Agriculture	−10.3	−12.4	6.4
Industry	−3.4	4.6	14.1
Transport	26.2	12.6	23.0
Construction	6.5	26.7	34.6
Others	11.3	1.6	−8.2
Total	−2.9	2.9	13.8

SOURCE *China Agricultural Yearbook*, 1980.

output of these enterprises is estimated at 56.6 billion yuan.)[9] By the end of June 1981, these enterprises had a work-force of 30 million, and accounted for 31.7 per cent of the fixed assets of communes, brigades and production teams and 34 per cent of the total revenues for farming.[10]

From 1977 to 1979, the cumulative profits of the commune and brigade enterprises amounted to 23 440 million yuan, of which 6750 million yuan (or 28.8 per cent) were spent on supporting capital construction in agriculture (for example, construction of water control works and purchase of farm machinery) and on assisting the poor brigades and teams. In 1980, the total profits earned by these enterprises were 11 800 million yuan (compared to 10 400 million yuan for 1979) (see Appendix I, Table A.7) of which 2300 million yuan (or 20 per cent) were spent on agriculture.[11] Thus apart from creating direct employment, these enterprises also contributed to employment generation indirectly by providing capital for agriculture.

Table 6.2 above shows that considerable inter-sectoral shifts have taken place between 1978 and 1979 in commune and brigade enterprises. The number of agricultural and industrial enterprises has declined while that of transport and construction enterprises has increased. Enterprises in these two sectors also account for the highest increases in output and employment.

Most of the information on employment, gross output, profits and fixed assets (see above and Appendix I, Table A.7) relates to all the commune and brigade enterprises, and not to rural industrial enterprises alone. Separate data for these latter enterprises are not available to us. However, we assume that that the information presented here is also representative of the rural industrial enterprises, which account for a little less than 50 per cent of the total number of commune and brigade enterprises in the province of Hunan (see Table 6.3).

In 1980, Hunan had 103 000 commune and brigade enterprises generating an output of 3100 million yuan or 31 per cent of the total commune output, and employing over 1.5 million persons, or 8 per cent of the total labour force.

Table 6.3 shows that between 1976 and 1980, the total number of commune and brigade enterprises in Hunan fell by 14 per cent and their employment by 18.5 per cent. However, the total gross output of these enterprises increased considerably. It seems likely that the smaller labour-intensive and less profitable enterprises were shut down. For example, it is estimated that 3100 commune and brigade enterprises in Hunan were closed in 1980, while more than 4000 were merged or transferred to other types of production.[12] Since 1979, attempts have been made to merge very small enterprises into larger ones in order to make them more efficient.

Comparison between enterprises operated by the communes and those run by brigades shows a considerable difference (about 53 per cent) in the output per worker. This difference may be due to the much higher labour productivity in the commune enterprises, or it may also be partly explained by the differences in the product-mix and the selling prices of fixed products in the two types of enterprises. However, there is one difficulty in interpreting the data in Table 6.3. The value of output is gross rather than net value added. If the brigade enterprises produce semi-processed goods for the commune enterprises, the former's output might be much lower than the latter's. Unfortunately we have no data on net value added for the two types of enterprises to determine whether this factor explains the large difference in output per worker.

A comparison between Tables 6.2 and 6.3 shows that the total number of commune and brigade enterprises declined in both Hunan and China as a whole. However, the Hunan pattern deviates somewhat from the national pattern. For example, in the case of China, while the number of enterprises declined, both output and employment increased whereas, in the case of Hunan, the decline in the number is associated also with a decline in employment.

Hunan's experience is compared with that of other provinces of the Central–South Region in Table 6.4 below. Gross output per enterprise is very similar in Guangxi, Hunan and Hubei whereas it is much

TABLE 6.3 *Output and employment in commune and brigade enterprises in Hunan Province (1976–80)*

Item	*All enterprises**		*% change*	*Industrial enterprises*		*Non–industrial enterprises*
	(1976)	*(1980)*		*Commune (1980)*	*Brigade (1980)*	*(1980)*
Number (000)	120	103	–14	12	33	58
Total employment (000)	1 900	1 530	–18.5	450	460	620
Total output (million yuan)	1 580	3 100	100†	1 200	800	1 100
Total value of fixed assets (million yuan)	–	2 250	–			
Output per worker (*yuan*)	0.8	2.0		2.6	1.7	1.8
Employment per enterprise	15.8	14.9		37.5	14.0	10.7

*All enterprises include agricultural, industrial, construction, transport and service enterprises.
†This percentage may be misleading since it is not clear whether total output is at current or constant prices for the two dates.

SOURCES *Provincial Bureau of Commune and Brigade Enterprises*, Changsha, Hunan, and *China Agricultural Yearbook*, 1980.

TABLE 6.4 *Output of commune and brigade enterprises in the Central–South Region (1978–80)*

Province	*Output per enterprise* (1979) (000 yuan)*	*Employment per enterprise† (1979)*	*Agricultural output of brigade enterprises (1979) (100 million yuan)*	*Industrial output of enterprises (100 million yuan)*				*Percentage change in industrial output*	
				Communes (1978)	*Communes (1979)*	*Brigades (1978)*	*Brigades (1979)*	*Communes*	*Brigades*
Henan	1.7	22.1	15.46	11.82	11.50	12.66	18.07	–2.7	42.7
Hubei	1.2	13.8	7.69	8.06	8.54	4.27	4.66	5.9	9.1
Hunan	1.3	16.5	7.65	10.30	10.97	6.42	7.99	6.5	24.4
Guangxi	1.2	19.1	3.68	3.72	3.75	2.0	2.05	0.8	2.5
Guangdong	1.6	23.2	6.96	13.40	13.97	9.98	10.64	4.2	6.6
Central–South Region	1.4	18.3	41.44	47.30	48.73	33.53	43.41	3.0	29.4
Total China	1.7	19.6	197.98	224.11	239.17	161.15	183.67	6.7	14.0

* Derived from Appendix I, Table A.6.
† Derived from Appendix I, Table A.6.

SOURCES *China Agricultural Yearbook*, 1980, and Bureau of Commune and Brigade Enterprises, Ministry of Agriculture.

higher in both Guangdong and Henan. Employment-intensity of these enterprises is one of the lowest in Hunan, and the highest in Guangdong followed by Henan. This suggests that the enterprises in these two latter provinces are larger, labour-intensive and more productive than those in Hubei or Hunan. In Hunan, the brigade industrial enterprises contributed less output than the commune enterprises. This is also true of Guangxi, Guangdong and Hubei. But the reverse is the case in Henan where industrial brigade enterprises are relatively more important. During 1978–9, their output increased by 42.7 per cent while that of commune industrial enterprises actually declined. In all the five provinces as well as China as a whole, the industrial brigade enterprises registered a much larger increase in output than the commune enterprises during 1978–9.

INPUTS, FINANCE AND ORGANISATION

A rapid growth of rural small industry in Hunan has been hampered in the past due to a number of factors. First, there were very few market surveys to determine the demand potential for rural industry products. In the absence of such surveys and feasibility studies, a number of inefficient plants were established. Second, small-scale and large-scale plants have been competing for scarce raw materials supplies.[13] Third, all rural plants were not covered by the provincial plan which meant an imbalance between production and market needs. Fourth, there was no unified planning and distribution of materials and products between town and county, large and small industry, and between heavy and light industry. Finally, the quality of products is not standardised and the methods of production are often outdated, yielding very low capital and labour productivity. This section addresses itself to some of these questions as well as to those related to the financing and administrative organisation responsible for rural small industry at the provincial level.

Labour Supply and Training

Most rural enterprises in Hunan province are staffed by local labour from the production brigades, for example the bicycle factory outside Changsha and the porcelain factory in Zhu Zhou that we visited (see Appendix II). Labour works either full-time or seasonally during agricultural slacks. Since farm labour is paid for by the production brigades and teams to which it belongs, problems of foodgrains shortage and of urbanisaton of the labour force are avoided.[14] Furthermore, since the workers are from the farm, they are accommodated there thus saving on costs of accommodation and other related expenses. This facilitates additional employment generation at low cost.

The supply of agricultural labour to the rural industrial enterprises has increased since the early 1960s when no more than 2–3 per cent of the labour force of the brigades and teams could be used for rural non-farm activities (presumably, this measure was introduced to overcome a temporary shortage of the agricultural labour force during the Great Leap Forward period). However, as we noted in the last section, in 1979, the commune and brigade enterprises in China accounted for 29 million or 9.4 per cent of the total agricultural labour force, and in 1980, 30 million or 10 per cent. During our trip to Hunan in November 1981, the Chinese officials mentioned that an estimated 30 per cent of the agricultural labour force could be released as surplus. Despite this large size of rural underemployment in China, there are at present no 'unlimited supplies of labour' to the urban sector since the ratio of marketable to total output of foodgrains is too low for the urban sector to sustain the speed of industrialisation (and hence of urbanisation) resulting in an increase in the ratio of urban to rural population.[15] It is therefore understandable that the Chinese government is determined to stem the migration of rural labour to the urban areas by expanding rural commune and brigade enterprises.[16]

One of the most important benefits of the Chinese rural industry is its contribution to skill formation through 'learning by doing'. At the commune, brigade and team levels, agricultural workers undergo apprenticeship training in such rural industrial activities as repair of farm tools, water pumps, electric motors and so on.

The Hunan province has a number of training facilities for skill formation in rural industry. These facilities are of both a formal and informal nature. For example, three types of training are mainly imparted at the county level. These are: (i) training of accountants; (ii) training of teachers who, once trained, teach at the commune level; and (iii) training of technicians. However, most of the technician training takes place at the provincial or district level.[17] On the other hand, less specialised training for the manufacture of firecrackers, for instance, is done mostly at the county level. Training in the operation of agricultural machinery is given at the Hunan provincial technical training school.

Short courses of management are organised for heads of bureaus of agriculture and for commune and brigade enterprises at the provincial and county levels of administration in Hunan. This is a research-type course that indirectly helps the rural enterprises which are the administrative responsibility of these bureaus.

A number of facilities also exist for the training of workers in small industrial enterprises. First, workers in rural enterprises are sent periodically to educational institutions (secondary schools) for training for a period of 1–3 months. Second, training is also provided to rural industries in big factories which subcontract work to rural industrial enterprises; this facility is normally used for training new workers. Third, training within factories is provided in evening classes which take place twice a week. Fourth, middle-level and secondary schools have been established specifically for imparting vocational and trade skills to rural enterprise workers. This training lasts for a period of two years.

Supply of Raw Materials and Equipment

The physical resources like raw materials and equipment are generally supplied locally from within the commune structure. In fact, the old policy of rural industrialisation emphasised the use of local waste and scrap material not required by large industry and thus having a negligible opportunity cost.[18] This local self-reliance is still being encouraged except that recent measures of economic decentralisation and the introduction of market mechanism have led to competition between rural and urban industry for raw materials. For example, in the province of Hunan small leather tanneries and paper mills are known to compete with more efficient larger mills for raw materials. The smaller enterprises are also alleged to pollute the environment.

Some critics argue that the competition between rural and urban industry for raw materials and markets is becoming severe and that the small enterprises are relatively inefficient thus leading to a waste of scarce resources. According to Li Yu and Chen Shengchang:

> Rural communes and production brigades set up coal pits next to big coal mines. Their rate of recovery at 10 per cent is nearly 40 per cent lower than that of big mines. Competition between small coal pits and big coal mines is so intense (in the Rujigou district in Helanshau Ningxia Hui Autonomous Region) that in many places

> big coal mines are unable to operate . . . Similar conditions exist in Hunan, Hebei, Henan, Heilongjiang, Inner Mongolia and Xinjiang.[19]

The Hunan government officials that we interviewed in November 1981 were also very preoccupied by the problem of the relative inefficiency of rural small enterprises in a number of industries. We were told of the decision of the provincial government to gradually transform the small-scale paper mills into medium- and large-scale mills with an annual capacity of 10 000 tons of paper.[20] Similarly, haphazard development of rural small tanneries which competed with larger ones for raw materials was to be prevented through government intervention.

In order to curb competition between rural and urban industry, the Chinese government urges the latter to subcontract work to the former on an increasing scale. There are some examples of urban industry transferring the production of spare parts and other goods to rural commune and brigade enterprises. The brigade-operated porcelain factory in Zhu Zhou (see Appendix II) initially worked on a subcontracting basis for a large factory in the city. The parent factory provided technical assistance in product design and marketing. Under the subcontract, the parent company also provided technical assistance in management and training. However, it did not supply any equipment which was purchased by the brigade factory independently out of its own savings. The state regulations concerning commune and brigade enterprises enforced in 1979 do provide for transfer of 'special equipment and tools at fixed prices' by the urban industries.[21]

There are indications that the policy of subcontracting to promote cooperation between rural small industry and urban large industry is not working too well owing to reluctance on the part of the urban industries. The latter feel that 'spreading out production involves problems, equipment, material and capital . . . it is difficult to regularly supply the commune and brigade factories with raw materials, provide technical guidance in production and inspect the quality of their products'.[22] The urban-based industries seem to be afraid that they might not be able to fulfil their own quotas if the commune and brigade factories failed to guarantee product quality.[23] This seems to be a genuine problem which is implicitly recognised in the state regulations (noted above) in the following words:

For cooperation between urban industries and rural people's communes and brigades to be maintained, supplies for raw and semi-finished materials and marketing channels must not be interrupted. Commune and brigade-run enterprises must guarantee the quality of products and fulfil their assumed tasks on time.[24]

Competition for scarce raw materials can also be checked by the state allocating such materials as steel, cement and wood on a planned basis to both rural and urban industry. The rural enterprises which are incorporated into the state plans receive materials and equipment through the county (*xian*) bureau of commune and brigade enterprises. Under the new state regulations, the local planning commissions at various levels are expected to assess the materials required by the commune and brigade enterprises for production, maintenance of equipment and technical improvements. In the light of these assessments, the State Planning Commission is to make the necessary allocations for these enterprises.

We were told that in the past these government allocations were always below the requirements of the communes and brigades. However, since the economic reforms in 1979, the commune and brigade enterprises are guaranteed allocations of raw materials and equipment to satisfy their requirements fully. Cutbacks on heavy industry have released raw materials like steel for rural industrial enterprises. The new measures also stipulate that 'a certain amount of sub-standard steels and other waste materials recovered by the localities is to be allocated by the local planning commissions to commune and brigade enterprises'. In allocating raw materials, the state gives higher priority to poorer communes and brigades, especially in the areas inhabited by the minority communities. Finally, in some cases, (such as textiles) the state policy of raising procurement prices for raw materials offers an incentive for the provision of adequate supplies of raw material inputs.[25]

A large number of commune and brigade enterprises in Hunan are not covered by the state plan. This means that these enterprises have to obtain raw materials and equipment locally from the market, from their large-scale parent enterprises which may subcontract, or directly from the departments or supply companies within Hunan or in another province.

The materials that are supplied by the state within the plan framework have to be bought at the state-fixed prices. However, materials obtained outside the plan (say in the open market) can be bought at wholesale prices. The departments concerned with rural industrial enterprises are expected to make available appropriate transportation for raw and semi-finished materials.

Financing

The financial resources required by Hunan's rural industry come largely from within the communes or brigades. It is estimated that financing of up to 80–90 per cent of the total expenditure is provided by the savings of the farm families, brigades and communes, with the remainder being supplied by loans from industry, the government or the Farmers' Bank.

The provincial and central governments are not very important sources of funds for the commune and brigade enterprises. However, normally not less than half of the investment made by the state to support rural communes must be allocated to the responsible departments for assistance to poorer brigades to establish rural enterprises. The departments generally entrust these funds to the local banks for disbursements to the communes and brigades. Loans are generally granted at low rates of interest (about 1.7 per cent per annum). It is reported that loans which are spent on buying equipment are generally repayable over a period of 3–5 years. The more general loans may also be repayable over a longer period depending on the levels of income of the borrowing communes and brigades.

At present, the grant of credit to Hunan's rural industry is very limited despite the fact that the lack of finance is reported to be one of the major problems in its promotion. There is no government policy against the grant of adequate credit to rural enterprises. Nevertheless it appears that the communes and brigades prefer to limit their debts, especially since they have been accustomed for so long to relying on their own savings.

Since 1980, a new policy has been adopted in Hunan under which communes and the individual members from communes and/or brigades can jointly own and manage rural enterprises. A member is entitled to buy a share on which he or she gets an interest of 15 per cent in the first year and a little less in the subsequent years. There is no formal limit to the number of shares that can be bought but, in practice, current low household savings limit ownerships. The shareholders form a standing committee which supervises the management of the enterprise. Profits are distributed in proportion to the shares in the enterprise.[26] Usually the shareholding members own up to 70 per cent of the enterprise with the remaining 30 per cent owned by the commune or the brigade.

In some cases, enterprises are jointly managed by large urban industry and the rural commune/brigade industry. For example, capital may be shared equally between the two. To date, however, large industry makes only a minor contribution to the financing of rural industry.

The promotion of sideline production is seen as a means to accumulate local savings for investment in rural industry. To give an example, milk-producing factories in Hunan receive milk supplies directly from individual families rather than from the communes. This offers a private source of income to the households.

In Changsha, we were told that the Hunan rural industry suffered from severe financial constraints, since support to commune and brigade enterprises from the national and local budgets was almost negligible.[27] The commune and brigade level accumulation funds had to be allocated to workers' welfare facilities besides being invested into new enterprises or expansion of existing ones. Thus they were inadequate for financing rural industry.

Taxes and Profits

A general policy of tax relief for commune and brigade enterprises should also indirectly raise the potential resources of communes and brigades for investment in rural industry. The state has introduced reduced tax or tax exemptions for rural enterprises as one of the measures for their promotion. The enterprises in the better-off regions are exempted from taxation till the profit rate reaches 3000 yuan a year – a level that is much higher than the earlier level of 600 yuan a year. The enterprises which directly serve agricultural production and which are responsible for the livelihood of most of the commune and brigade members are entitled 'to register specific products and services for exemption from industrial, commercial and income taxes'.[28]

Since 1978, small iron mines, coal fields, power stations and cement works have been exempted from taxes for a period of three years. In order to promote the growth of income-earning opportunities through rural industry in the poorer communes and brigades (for example in autonomous regions inhabited by minority nationalities), from 1979 onwards all rural enterprises in the border areas were granted a general tax holiday for five years.[29]

In general, commune and brigade enterprises pay tax to the state at the rate of 20 per cent. The same rate applies to the supply and marketing companies of the bureaus or departments in charge of these enterprises. These enterprises do not transfer any portion of their profits to the state. However, there are general guidelines set by the state for the use of these profits. It is stipulated that

> profits must be expended mainly on farm capital construction, the mechanisation of agriculture and in giving support to the poor teams. A certain proportion may also be distributed among the commune members or spent on commune and brigade welfare undertakings. Waste and extravagance are prohibited and it is forbidden to spend profits on entertainment or gifts.[30]

The commune and brigade enterprises are also entitled to 'the shares of the profits made from the sale of export commodities in foreign currency in accordance with the unified regulations of the state'.[31]

Administrative Organisation

One of the main provincial bureaus dealing with rural industry in Hunan is the Bureau of Commune and Brigade Enterprises (CBE). This bureau, of recent origin (established in 1977), is divided into three departments and a company or corporation as follows: Production and Technology, Management, Finance, and a supply and sales company. The company also handles information exchange. The Bureau is under the responsibility of the Office of Industry and Transport.

The CBE Bureau plays an important role in the procurement of production orders for commune/brigade industries. For this purpose, it surveys production requirements of county factories and commercial corporations and matches the productive capacity of the commune enterprises with them. The Bureau also provides supplies of raw materials for the rural industrial enterprises through the supply corporation. The latter has to apply to the provincial and county planning committees for the allocation and procurement of these materials. This assistance by the Bureau applies only to those cases where commune enterprises produce either for 'parent' state-owned enterprises at the provincial or county level, or for the export markets. In the case of production for limited local markets of the communes or brigades, the enterprises procure their own supplies directly from state factories with surplus materials supplies.

It is clear from the foregoing that part of the total output of the rural enterprises falls within the purview of state planning and is thus subject to the control and scrutiny of a number of bureaus at the county and provincial levels. For example, the Committees of Economic Reconstruction and Planning are required to undertake appraisals of requests from rural commune enterprises for raw materials supplies in the light of which allocations are made. The Bureau of Foreign Trade and its corporations may be required to negotiate quotas for production supplies from the commune enterprises destined for the export markets. Similarly, the Bureau of Agricultural Machinery may be involved in the approval and supply of agricultural machinery required for farm mechanisation and for the manufacture, repair and maintenance of equipment.

In addition to the county bureaus, each group of industries is administered by a bureau at the provincial level and a ministry at the central level. Thus the CBE Bureau shares responsibility with the bureaus and ministries dealing with particular groups of industries.

Apart from the administrative organisations discussed above, there are a number of research institutes and information centres dealing with specific industries; for example, the Tea Information Centre which deals with tea cultivation comes under the National Research Bureau, whereas a second Tea Information Centre deals with processing machinery and is the responsibility of Foreign Trade.

CREATION OF NEW ENTERPRISES

The CBE Bureau is responsible for the approval or rejection of applications for the establishment of new enterprises. Any member of a commune is free to propose the establishment of a new enterprise to the commune director/manager who in turn may refer the matter to the CBE Bureau at the provincial level. The Bureau will examine whether the proposed enterprise duplicates and competes with some existing enterprises in the state and collective sectors. The production and technology department of the Bureau will investigate the commune's requirements for the construction and production materials, energy, skilled manpower and other facilities, besides considering its industrial experience and composition of its production. The Bureau will also undertake a market survey. If it is satisfied that the new enterprise will cope with unmet demand, it may give approval for its establishment and refer the matter to the Planning Committee and the Committee of Economic Reconstruction. These provincial bureaus will undertake a planning and economic feasibility study: they may also consult a sectoral (industrial) bureau depending on the nature of the industry in which the enterprise is proposed to be established.

Approval of the Hunan Provincial Planning Commission is required only if the proposed enterprise requires investment of over 50 000 yuan. For enterprises requiring lower investments, the district or county-level planning committees (bureaus) are authorised to take final decisions.

Once the proposed enterprise is established it may require equipment and technology from abroad. The commune responsible for the enterprise will have to go through the Trading Company in the province to obtain equipment. The Hunan Bureaus of Foreign Trade and Machine Building Industry will also be involved, inter alia, with the situation of the enterprise in respect of raw material supplies, availability of local skills, market potential for the product, and so on, before a final decision is taken to permit import of technology.

The foregoing discussion amply demonstrates the sharing of responsibility among a large number of administrative units in the control, management and operation of rural enterprises. Involvement of several committees and bureaus in making a single decision is likely to lead to bureaucratic red tape and duplication of effort. There is thus a need for an overall coordination machinery. Neither the CBE nor the Committees of Planning and Economic Reconstruction seem to perform such a function for an integrated development of rural industry in Hunan.

RETROSPECT AND PROSPECT

China has a long but varied experience with rural industrialisation which can be grouped under the following historical epochs, viz. the Great Leap period (1958–60), the Cultural Revolution period (the late 1960s) and the post-Mao period (1976 onwards). The first epoch was perhaps the beginning of a large-scale introduction of rural small-scale industries within the framework of the policy of 'walking on two legs'.[32] This planned industrial and technological dualism was not an unmixed blessing for the growth of efficient rural industries but it offered some positive learning effects. By the early 1960s many of the rural local industries established during the Great Leap Forward period were already dismantled or merged into larger enterprises. However, in spite of the political vicissitudes and shifts in policies during the different epochs, rural industries have remained important in the rural development of China. This is particularly true in respect of their contribution to agriculture through the provision of inputs and to the total industrial output. It is reported that commune and brigade enterprises in China account for 15 per cent of the total industrial output of the country. The output of the brigade enterprises (which are smaller in scale than the commune enterprises) is estimated at about 50 per cent of the combined output of the commune and brigade enterprises.[33]

The continued importance of the rural industry is attributed to a number of factors, namely transportation bottlenecks and the resulting segmentation of product and factor markets, dependence on the policy of self-reliance and balanced regional development, and the application of appropriate technology using local materials and skills. The shifts in development policies from one epoch to the other did not fundamentally alter the nature and role of the above factors in explaining the importance of rural industries in China.

The importance and growth of Chinese rural industry cannot be explained in terms of pure economic factors. A number of ideological factors have also been quite favourable to the growth of rural industry in the past.

Mao's egalitarian ideals and attitude of land ownership must have given support to rural industry.[34] Unlike most other developing countries, the similarity of life styles between the élites and the masses ensured during the Mao regime kept in check the demonstration effect and the resulting luxury consumption of products mostly imported from abroad. Furthermore restrictions on labour movement from urban to rural areas, and vice versa, also prevent the demonstration effect from reaching the rural areas. However, it is not clear whether these arguments in favour of rural industry still hold valid in the post-Mao period with considerable international economic and political relations with the West. The availability and use of a variety of imported goods (Coca Cola and Pepsi Cola are just one example) strikes the eye of any visitor to China today. China can no longer be quoted as a suitable example of an economy which gives priority to products of domestic origin which are more appropriate to the factor endowments and life style of the country. Today in China, a demonstration effect matched by the local purchasing power might well lead to the import and production of goods that compete with the rural industry products, and which are not necessarily most suited to the Chinese economy.

Some authors believe that the contribution of Chinese rural industry to employment creation in the past was not as great as might have been expected. For example, Sigurdson estimated that in Hebei

Province in 1973, rural industries employed only 5 per cent of the labour force at the seasonal peaks.[35] The Americal Rural Industries Delegation quotes a figure of 4–8 per cent. Both Sigurdson and the American Delegation seem to underestimate the contribution of rural industrialisation to employment generation, since their reference period was 1973–4 or earlier when the rural industries were not allowed to absorb the large agricultural work-force. Although these industries were resurrected in the late 1960s and early 1970s, their reconstruction and growth was quite slow during the Cultural Revolution, and became steady only after 1970.

In Hunan province in 1980, rural industries employed 8 per cent of the total labour force (see Table 6.3). For China as a whole, over 30 million persons were employed in the commune and brigade enterprises (or nearly 10 per cent of the total employment in the commune sector). This indicates that these enterprises play an important role in employment generation in today's China.

It is useful to remember that the estimates for 1973 and 1975 were made at a time when the Chinese planners guaranteed full employment in the Socialist economy. The problem was seen essentially in terms of lower productivity and labour underutilisation. Since 1979 China has officially recognised open unemployment in the country and has abandoned the policy of guaranteed employment to every able-bodied person. In the light of this new policy framework, the Chinese authorities have explicitly emphasised the importance of rural industrialisation and the small urban 'informal sector' in China's new employment strategy. The future prospects for growth of Hunan's rural industries and their contribution to employment will depend, inter alia, on the government's attitude and policy towards the commune and brigade enterprises vis-à-vis larger urban enterprises. We were told in November 1981 that Hunan's official rural industrialisation policy is guided by the need to create additional employment and save scarce foreign exchange. This is further underlined by the State Council in the following statement on the significance of the commune and brigade enterprises:

> At the same time, it [the expansion of commune and brigade enterprises] will open up new avenues of employment in production for the labour power which will be released by mechanisation, make fuller use of local resources, aid the development of a diversified rural economy, increase collective income and raise the living standards of commune members.[36]

7 Concluding Remarks

We have covered a rather vast ground in this book in the temptation not to lose any information collected or experience gained during our interviews and field trips in the province of Hunan. A short stay in a country as vast as China hardly makes it possible to draw any meaningful conclusions. Nor does the rapidly changing character of Hunan and the Chinese economy.

The new economic policies were introduced in China no more than three to four years ago. This time span is too short for these policies to have achieved the desired results even if they had been fully and consistently implemented. There are indications that some of these policies have been modified whereas the implementation of others has been slowed down in the light of hindsight and the experience of preliminary results. The Chinese planners and policy-makers are known for their pragmatic and 'trial and error' approach to the handling of economic problems. The Chinese economy walks on two legs; it also walks two steps forward and one step backwards. This approach has its advantages. It introduces built-in flexibility to the economic system, which makes rapid adjustments possible. But it has its drawbacks too. Frequent changes in policy have a destabilising influence which makes their objective assessment almost impossible.

In the case of Hunan, several issues seem to be crucial (and also somewhat controversial) in its economic readjustment and transition to modernisation and economic growth. For example, the issue of competition between large urban industry and rural industry remains to be settled. Some recent surveys have shown that rural industry operated by communes and brigades tends to be inefficient. Second, the drive for export promotion has become a major policy issue since foreign exchange earnings are needed to pay for major imports of capital goods, intermediate products, and technical know-how. Unutilised industrial capacity is being utilised through heavy imports of various kinds of inputs needed for production. This drive for rapid export promotion through costly imports may not necessarily be consistent with the declared policy of employment generation. We shall make a few brief remarks on each of these issues below.

COMPETITION BETWEEN URBAN AND RURAL INDUSTRY

During the Mao period, the main rationale for the growth of rural industry was based on the utilisation of local resources and scrap and waste material from large factories which could not otherwise be utilised. Rural industry was not supposed to compete with large-scale urban industry for raw materials and skilled manpower. It was to develop under a production plan or materials distribution plan which meant that its expansion would occur only after the state procurement targets were fulfilled.

The rural industry flourished during the Great Leap period. During the 1960s and 1970s, the 'five small industries' and the 'commune and brigade industries' were allowed to grow on two conditions, viz.:

(i) that raw materials required would be supplied locally from the province, *xian* or commune, and not from the central government; *and*
(ii) that the products of the industry were to be locally distributed and were not allowed to be sold outside the locality.

During the phase of local industry development which began in 1970 or 1971, the CCP's Central Committee gave a directive that 'in order to solve the issue of slow agricultural growth, each locality should endeavour to establish industries surrounding agriculture and supply industrial inputs and materials to agriculture'. Thus self-reliance was the underlying rationale for the growth of local rural industry which was prevented not only from competing with urban industry for raw materials but also from competing with agriculture for the supply of manpower.

With an increase in the number and output of rural industry establishments, such inputs as power, materials and labour became increasingly scarce. The industry was forced to sell its output informally even outside the province in order to obtain raw materials in exchange. The extent of the materials shortage seems to have increased some time after the mid-1970s, thus leading to competition for centralised supplies of power and raw materials.

In a command economy with centralised control and management, it was perhaps easy to prevent competition between large urban state-owned industry and rural industry owned by the communes. The requirements of the latter for the state-controlled materials like high-grade iron ore and steel could be rationed or supplied on a planned basis. The problem of competition did not seem to come to the fore during the 1960s and early 1970s since it was kept under control through central intervention.

However, during the post-Mao period, greater flexibility and decentralisation of management at different levels seems to have accentuated the problem of competition between rural and urban industry. This may have been further aggravated by the official encouragement of rural local industry with a view to creating employment.

The current debate on rural industry in China is also concerned with its alleged relative inefficiency vis-à-vis large urban industry. The opponents of rural small industry claim that it incurs heavy losses, consumes large quantities of fuel, and makes an inefficient use of other raw materials. For example, it has been estimated that for smelting enterprises, the ratio of fuel (coke) consumption per furnace in 1978 was 950 kg for small plants, 763 kg for medium-sized plants and 562 kg for large plants.[1] The high costs for small plants might be due to diseconomies of scale.

Notwithstanding the above problems the rural small industry has survived and even flourished in Hunan in several activities. This may have been due in part to market segmentation and poor transportation facilities.

The planners in Hunan also seem to make a conscious effort to promote efficient small-scale production in rural areas. This seems to be indicated by the planned mergers of very small inefficient enterprises into larger economically viable ones. As indicated in Chapter 6, the total number of small factories in Hunan was reduced between 1976 and 1980. If this was the result of mergers alone, the average employment per enterprise would have increased. In fact, however, for 'all enterprises', employment went down from 15.8 to 14.9 on average (see Table 6.3). This suggests that many enterprises must have gone out of business, and others must have introduced labour-saving technology.[2]

The above point suggests that the promotion of rural industry in Hunan needs to be planned on the basis of both 'efficiency' (economic costs) and 'equity' (employment) criteria which can be in conflict at least in the short term. Following a pure efficiency objective is likely to sacrifice employment which is one of the important social objectives of the planners in Hunan. Even if rural industry competed with large urban industry and were relatively inefficient, it might still be socially desirable to promote its growth considering that the

Chinese planners of today attach a higher weight to additional employment and consumption than did their predecessors.

The rural industry, especially that part of it which competes with larger urban industry, may deserve an infant-industry type of protection on efficiency as well as equity grounds. Indeed, the Chinese planners offer a number of incentives for the growth of rural small industry. Furthermore the recent establishment of a special bureau of commune and brigade enterprises in each province is a clear indication of a high priority given to rural industry.

Competition between rural industry and urban large industry for raw materials and markets can be more easily checked within the planned economy of China than in a mixed economy like that of India where a somewhat similar debate took place in the 1950s concerning the efficiency of 'cottage' industries and their competition with large-scale organised industry.[3]

The officials in the Hunan Bureau of Commune and Brigade Enterprises need to identify specific rural industries which compete with urban industry for raw materials, and to assess their total requirements for these materials. In the light of such assessments, the planners in Hunan need to determine the share of each type of industry within the overall planned allocation of raw material resources. Thus the incorporation of rural industry into overall planning for industrial inputs and outputs, combined with the promotion of subcontracting by large urban industry to rural industry, should facilitate a complementary relationship between these two categories of industries. A positive linkage and consistency between local plans for rural industry and central plans for modern factory production become particularly important when rural industry produces goods for the latter under subcontracting arrangements.

EXPORTS *V*. EMPLOYMENT

Implementing the Chinese national policy, the Government of Hunan has considerably emphasised the expansion of exports and the use of foreign capital in its drive for rapid modernisation. Perhaps this shift in policy has been far too rapid for its capacity to absorb and pay for the import of capital and technology. Hence the recent slowing down, particularly in the import of some capital and foreign technology.

Hunan must of course expand its exports to pay for the needed import of foreign technology for its modern industry. A sound export promotion policy needs to tackle such issues as commodity composition and destination of exports, and the specific promotional measures needed to expand exports.[4] Since employment is one of the major objectives in Hunan as well as in China as a whole, the policy of export promotion needs to be consistent with this objective.

The factor endowment principle would suggest that the Chinese economy (including Hunan) should concentrate on labour-intensive products for export (assuming that there are foreign markets for them) since they embody its abundant factor labour and economise its scarce resources, such as capital and land. It appears that the factor endowment criterion does not explicitly guide Hunan's drive for export promotion. In November 1981, we found that many factories in Hunan, especially those with large excess capacity, were making a major bid to export its products (which were sometimes quite capital-intensive) which were being manufactured with expensive imported inputs. While full utilisation of existing productive capacity is desirable for expanding exports, it is doubtful whether it makes good economic sense to specialise in capital-intensive exports with a very high import-content.

An economic cost–benefit analysis is needed to determine the import-intensity of different exportable products. A comparison of import-intensity with their labour-intensity should provide useful guidelines for the formulation of Hunan's export promotion policy. In general, the export of capital-intensive and import-intensive goods would not be a sound policy. Nor would such a policy be consistent with employment promotion. It appears that the Chinese authorities now explicitly recognise the need to identify labour-intensive products for export (for example, foodstuffs, manufactures, and handicrafts) which may have a good world market potential.[5]

Appendices – Appendix 1 Statistical Tables *follow.*

APPENDIX I

STATISTICAL TABLES

TABLE A.1 *Total population of the Central–South Provinces of China (mid-year rounded to the nearest million)*

Province	*Total population (absolute figures and percentages)*										
	(1957)[1]		*(1973)*[1]		*(1979)*[2]		*(1981)*[2]		*(1982)*[3]		*Percentage increase (1957–82)*
	Total	*(%)*	*Total*	*(%)*	*Total*	*(%)*	*Total*	*(%)*	*Total*	*(%)*	
Henan	47.8	(7.4)	65.1	(7.2)	71.9	(7.4)	73.9	(7.4)	74.4	(7.4)	55.6
Hubei	30.4	(4.7)	42.1	(4.7)	46.3	(4.8)	47.4	(5.0)	47.8	(4.7)	57.2
Hunan	35.8	(5.5)	47.5	(5.3)	52.2	(5.4)	53.6	(5.4)	54.0	(5.3)	50.8
Guangxi	21.3	(3.3)	29.0	(3.2)	34.7	(3.6)	36.1	(3.6)	36.4	(3.6)	70.9
Guangdong	37.5	(5.8)	49.6	(5.5)	56.8	(5.8)	58.8	(5.9)	59.3	(5.9)	58.1
All China	646.5	(100.0)	–	–	971.0	(100.0)	996.2	(100)	1 008.1	(100)	55.9

SOURCES
[1] John S. Aird, *Population Estimates for the Provinces of the People's Republic of China*, 1953 to 1974, International Population Reports, No. 73 (Washington, D.C.: US Department of Commerce, February 1974).
[2] SSB, People's Republic of China.
[3] *Intercom*, vol. 10 (November–December 1982) no. 11–12.

TABLE A.2 *Output of major farm products of the Central–South Provinces*

Province	*Grain (100 million jin)*			*Other (10 000 dan)*		
	Rice	*Wheat*	*Corn*	*Cotton*	*Sugarcane*	*Tea*
Henan	40.9	216.7	96.1	710.1	102.7	3.1
Hubei	240.0	49.7	20.5	705.4	116.5	34.3
Hunan	399.7	5.3	4.2	187.6	1 784.2	128.5
Guangxi	196.3	0.4	24.0	1.3	10 549.8	16.9
Guangdong	295.8	2.8	1.3	–	27 397.0	27.2
Central–South Region	1 172.7	274.9	146.1	1 604.4	39 950.2	210.0
All China	2 879.1	1 192.8	1 184.1	5 935.2	59 336.2	658.2

SOURCE *Statistical Yearbook of China for 1981* (People's Republic of China: SSB, 1982).

TABLE A.3 *Average unit area yield of Major farm crops: Central–South China (1981) (at sown area)*

Province	*Grain*	*Cotton (kg/hectare)*	*Sugarcane*
Henan	2 565	555	32 198
Hubei	3 300	608	42 203
Hunan	4 005	548	52 253
Guangxi	2 925	225	41 535
Guangdong	3 270	–	62 040
All China	2 828	570	53 820

SOURCE *Statistical Yearbook of China for 1981.*

TABLE A.4 *Key agricultural indicators by provinces of the Central–South Region (1979)*

Province	*Indicator*					
	Arable area (million hectares)	*Irrigated area (million hectares)*	*Grain output (million tons)*	*Cropping index (%)*	*Distributed collective income per capita (yuan)*	*Gross value of agricultural output (billion yuan)**
Henan	7.14	3.63	21.34	152.9	63.4	10.29
Hubei	3.75	2.35	18.50	207.1	106.2	9.41
Hunan	3.44	2.44	22.18	242.3	92.3	9.41
Guangxi	2.62	1.46	11.73	194.2	74.7	8.11
Guangdong	3.22	2.15	17.38	217.4	88.4	4.82
Central–South Region	20.18	12.05	91.13	–	–	42.06
All China	99.49	45.00	332.12	149.2	83.4	158.43

* At 1970 prices.

SOURCE Information supplied by the State Agricultural Commission.

TABLE A.5 *Mechanisation: farm machinery stock in Central–South Provinces (1979–81)*

Province	*Machines*									
	Tractors (large and medium) (000)		*Hand tractors (000)*		*Irrigation machinery (000 hp)*		*Grain-processing machines (000 units)*		*Threshers (000)*	*Combine-harvesters (sets)*
	(1979)	*(1981)*	*(1979)*	*(1981)*	*(1979)*	*(1981)*	*(1979)*	*(1981)*	*(1981)*	*(1981)*
Henan	52	65	106	150	7 819	7 930	307	310	122	938
Hubei	33	37	98	114	2 705	2 928	206	225	177	658
Hunan	18	20	56	71	3 004	3 935	147	171	32	264
Guangxi	22	22	89	108	892	1 000	101	114	73	713
Guangdong	20	20	119	140	2 358	2 320	79	84	133	259
Central–South Region	145	164	468	583	16 778	18 113	840	904	537	2 832
All China	667	792	1 671	2 037	71 221	74 983	2 912	3 194	2 517	31 268

SOURCES For 1979, State Agricultural Commission and Bureau for Commune and Brigade Enterprises, Ministry of Agriculture; For 1981, *Statistical Yearbook of China for 1981*.

TABLE A.6 *Data on commune and brigade enterprises by provinces of the Central–South Region (1979)*

Province	*No. of communes*	*No. of brigades*	*No. of production teams (000)*	*Commune households (000)*	*No. of enterprises (000)*	*Employees (000)*	*Gross output* (billion yuan)*
Henan	2059	43 121	375	13 450	70	1 551	2.649
Hubei	1 256	30 425	238	8 030	114	1 576	1.957
Hunan	3 304	46 378	420	10 630	126	2 080	2.842
Guangxi	972	12 738	217	5 910	37	708	0.841
Guangdong	1 927	26 173	359	9 650	89	2 067	3.344
Central–South Region	9 518	158 835	1 609	47 670	436	7 982	11.633
All China	53 353	698 613	5 514	174 910	1 481	29 093	49.110

* At 1970 prices.

SOURCE Bureau of Commune and Brigade Enterprises, Ministry of Agriculture.

TABLE A.7 *Commune and brigade enterprises: economic data (1976–9)*

Item	*Year*		*Percentage change (%)*
	1976	*1979*	
Number of enterprises (000)	1 110	1 480	33.3
Employment (000)	17 900	29 000	62
Gross output (billion yuan)	27.2	49.1	80.5
Total profit (billion yuan)	7.8	10.4	33.3
Fixed assets (billion yuan)	17.5	28.0	60
Gross output as percentage of national total	23.3	30.6	

SOURCE *China Agricultural Yearbook,* 1980, State Agricultural Commission.

TABLE A.8 *Gross value of industrial output by provinces of the Central–South Region (1957–79) (millions of 1957 yuan)*

Province	Year												
	1957	*1964*	*1965*	*1966*	*1968*	*1969*	*1970*	*1971*	*1972*	*1973*	*1974*	*1975*	*1979**
Henan	1 560	2 886	3 737	4 844	0	5 476	7 173	8 456	9 717	0	10 612	11 960	19 748
Hubei	2 514	3 404	4 140	5 133	3 704	5 000	7 500	8 738	0	0	9 238	0	21 894
Hunan	1 634	2 301	2 929	3 705	3 072	4 531	6 027	6 509	7 029	7 901	8 167	8 403	18 178
Guangxi	717	927	1 307	1 633	0	1 970	2 426	3 014	3 518	4 075	4 520	4 926	10 864
Guangdong	3 570	6 441	7 810	9 305	7 678	10 596	12 553	13 292	14 324	16 412	17 474	20 284	24 599
All China	70 400	117 232	143 918	175 619	152 874	199 078	242 176	268 188	293 136	325 955	345 833	378 480	534 377

* Data for 1979 are from Chinese official sources (SSB) and are converted into 1957 yuan.

SOURCE R. Field, N. Lardy and J. Emerson, *Provincial Industrial Output in the People's Republic of China, 1957–75* (Washington D.C.: US Department of Commerce, 1976) p. 11.

TABLE A.9 *Investment in large, medium and small construction projects: Central–South Region (1981)*

Province	*Investment (100 million yuan)*		*Share of total investment (%)*	
	Large and medium projects	*Small projects*	*Large and medium projects*	*Small projects*
Henan	9.38	9.60	49.4	50.6
Hubei	8.29	12.77	39.4	60.6
Hunan	2.22	10.25	17.8	82.2
Guangxi	1.82	5.01	26.6	73.4
Guangdong	5.52	26.27	17.4	82.6
All China	169.18	252.55	40.1	59.9

SOURCE *Statistical Yearbook of China for 1981*, p. 315.

APPENDIX II
NOTES ON VISITS TO INDUSTRIAL ENTERPRISES AND COMMUNES

In November 1981, we collected some primary data in Hunan province on its small rural enterprises and large industrial enterprises. This appendix gives notes of our visits which, though not rigorous, give some indications of Hunan's industry.

HUNAN'S RURAL INDUSTRY

Porcelain Factory (Zhu Zhou)

It is a brigade-operated enterprise which started production in 1978. It produces 2 million sets a year for both domestic and foreign markets. Seventy per cent of its total output is exported to Canada, Iran and the United States through the Hunan Provincial Import–Export Corporation. The brigade deals directly with the corporation without going through the commune. It has direct contact with six to seven wholesale buyers in the province.

The factory produced nine varieties of products until 1980 when two more varieties (flower vase and wine containers) were added. On average, the factory price per piece ranges from 0.40 to 0.60 US dollars depending on the nature of the article. Diversification of its product–mix suggests a growing demand for its products. The total annual output of the factory is valued at 800 000 yuan or 28 per cent of the total output. The factory has plans to expand production by reinvesting profits. The brigade does not receive any capital from the commune for its expansion. The initial capital for the expansion of the enterprise was financed out of the brigade's own savings. All income from the factory is retained by the brigade although it has to pay taxes to the commune. Apart from the payment of taxes, the brigade factory is largely independent of the commune. The brigade makes its own decision about the appointment of the factory manager without any interference from the higher level of authority.

The factory uses raw material (clay) that is partly available within its surroundings. Some of the clay has to be imported from two provincial cities.

The factory employs a full-time staff of 124 persons including five technicians and five administrative personnel. All the workers come from within the brigade. The technicians working in the brigade factory were trained by Zhu Zhou city.

The average monthly wage per worker is 39 yuan with an extra 7–9 yuan bonus based on production. The factory adopts a payment system under which workpoints are assigned to individual workers on the basis of hours worked and the quality of work performed. Ten points are given for one work-day (we were told that in other factories these points ranged from 6 to 12 depending on effort and skill).

TABLE A.10 *Basic data on a bicycle factory in Hunan (1981)*

Area	10 000m^2
Physical output (pieces)	30 000
Value of output (million yuan)	4.1
Equipment (pieces)	140
Number of workers employed	168
Wage per month (yuan)	48
Annual wage-bill (thousand yuan)	96.8
Profits (million yuan)	0.2
Retail price per bicycle (yuan)	143
Ex factory price (f.o.b.) (yuan)	135

SOURCE Data supplied by the factory.

The techniques of production used by the enterprise are quite labour-intensive at different production stages, as well as for labelling and packaging.

The main brick kilns for baking pottery are big and coal-based, whereas those for baking glazed cups after patterns are fixed on them are small in size and are operated by electricity.

Bicycle Factory (Changsha)

This factory is operated by a collective farm under the responsibility of the Ministry of Agricultural Reclamation. Originally, the factory produced lathes and other machine tools. Since 1980, it has been converted to produce bicycles which are in short supply throughout Hunan and elsewhere in China.

The basic economic data on the factory are given in Table A.10. Two surprising facts reported in this table are worthy of some discussion. First, the difference between the ex-factory price and the retail price of bicycle is very small, thus suggesting a disproportionately low cost of transport, marketing and distribution. We were told that the retail price of the factory's bicycle in Beijing and other distant places would be much higher than 143 yuan on account of transportation costs. Although the retail price, established by the state, is below the real scarcity price of bicycles, it still represents 2–3 months' wages for a factory worker and almost a year's income for a peasant.

Second, profits are more than twice the wage bill. Wages may be depressed in order to finance accumulation. As we noticed in Chapter 2, this may be particularly true under the commune and brigade organisational structure with deferred wage payments and/or non-wage modes of employment. The factory employs only 168 workers out of a population of 3000 on the farm, that is, less than 6 per cent of the total economically active population. It does not have to provide accommodation for workers since they are regular members of the farm. This may make low wage payments possible. Many of the workers in the factory are women who work full-time, their families being engaged in farming. Of the total profits, 60 per cent are retained by the factory whereas the remaining 40 per cent go to the collective farm and the state, presumably in equal proportions. Of the profits retained by the factory, 70 per cent are used for the expansion of production and 30 per cent for workers' welfare.

The fact that the factory is not very profitable is not surprising. We found that the labour productivity in the factory was quite low especially in the welding, paints and assembling departments. One worker assembled only three bicycles per day. The assembling process is extremely labour-intensive with three workers fitting spokes into the wheel simultaneously and getting in each other's way. However, in the parts workshop, labour productivity seemed reasonably high as was evidenced by signs of greater activity.

Several bicycle parts (for example, ball bearings, chains and flywheels) are produced within the factory whereas others are imported from other parts of China or from abroad. Bicycles are shipped in assembled form as well as in knocked-down shape.

There is a tremendous waste of space in the factory. The plant layout is quite bad. Different departments, such as painting, assembling and production workshops, are physically apart thus requiring the workers to walk long distances across open space. Over-capacity in building space is remarkably high in most Chinese factories. In the Hunan bicycle factory, excess capacity is explained by the fact that old buildings in the Collective Farm were converted first into a machine tool factory and then into a bicycle factory. The buildings were probably not originally designed as factories. This may further explain waste of space.

The workshop buildings are quite large, poorly lit and inadequately heated. No worker wore any safety glasses while welding or working on lathes. However, the factory building is fitted with ceiling fans for use in the summer months when it becomes quite hot.

The collective farm decided to shift production from lathes to bicycles since it was easier to sell bicycles than lathes. As bicycles are in serious shortage in China, the factory plans to expand their production further rather than diversifying into other products. The factory manager did not think it was difficult to double the factory output and sell it. Thirty thousand bicycles have already been sold since the beginning of 1981, and contracts for another 60 000 bicycles have been signed for 1982. By the end of 1983, it is planned to raise the output to 100 000 bicycles. At present, the factory produces about 4000 bicycles per month, or an annual output of 48 000 bicycles. The factory manager told us that the doubling of output in one year would not involve any bottlenecks in terms of supply of equipment, space or materials. It is, however, hard to believe that production could be doubled without any problems.

The factory does not produce any bicycle carts since there are several other manufacturing units in Changsha producing them. The factory manager indicated that they might instead manufacture some light farm tools and implements for which demand was rising.

Red Flag Commune (Xiantan)

This commune was established in 1963 by a group of rusticated educated youth from the city. It has a population of 2900 of which 1800 constitute the work-force. About 800 youths are school graduates from the city. The commune covers an area of 160 hectares of farm land, 94 hectares of orchards, over 200 hectares of tea plantations, over 400 hectares of paddy rice, 11 hectares of vegetable gardens, and 13 hectares under peanuts, water melons and industrial crops.

The commune has five agricultural units and 11 industrial enterprises, a number of building and transportation teams, one school and four shops. The rural small industrial enterprises cover mineral processing, canning, electrical appliances, starch-making, making of rice noodles, leather goods, tiles, and building materials. These factories have 500 pieces of equipment producing 30 types of products. Over the years, the commune has built more than 16000 square metres of houses/factory premises. The canning factory is one of the most recent industrial enterprises which was established in 1980. The Commune undertakes important industrial activity which contributes significantly to collective incomes.

During 1971–80, the total value of the commune's output increased by 23.1 per cent per year for both agriculture and industry. In 1981, the value of output was 8 million yuan, exceeding the target of 6 million yuan by 30 per cent. The 1981 output represented a 38 per cent increase over that for 1980.

The commune has a total accumulation fund of the order of 8 million yuan. The average income per person in the commune is 640 yuan per annum. In recent years, the standard of living of the workers and commune members has risen with the growth of farm output. This is indicated by the fact that there is now one bicycle for every three persons, and one wrist watch per person. Also half the number of the commune families have bought radio sets and sewing machines.

Embroidery Factory (Changsha)

This is a state-owned enterprise under the control of the municipal government of Changsha city. The factory manager is appointed by the municipal government.

Most of the factory employees are young and old women who work seven hours a day and six days per week. They get a break for rest every two hours, in order to relieve strain on their eyes. The average wage per worker is 50 yuan per month, or 600 yuan per year, which is a little below the average income per person in the Red Flag commune discussed above.

The factory has rather poor working conditions, particularly dim lighting, dampness and cold winds. The only ventilation of the buildings is through opening the windows which, in winter, exposes the workers to extremely cold winds.

The sewing department of the factory has both Chinese and imported sewing machines. The bulk of the imported machinery comes from Western Germany (mainly Pfaff sewing machines). We were told that productivity of the two types of machines is quite similar although the imported machines can operate a greater variety of more complicated and intricate designs.

There is a great shortage of sewing machines in China today. Next to bicycles, sewing machines are perhaps the most sought-after pieces of equipment for use in the households by female members. The Chinese women like to tailor their own clothes at home, hence the growing demand for sewing machines. Tailoring is also a lucrative sideline activity that can be done at home.

The local emboridery designs used by the factory are printed in the form of a booklet. The factory premises have a sales workshop which sells embroidered pieces of cloth as well as Chinese paintings.

HUNAN'S LARGE INDUSTRY

Industrial Pump Factory (Changsha)

This factory was established in 1951 as a state-owned enterprise under the control of the Bureau of Mechanical Industry of the Hunan Provincial Government. It covers an area of 600 square metres.

Output

The factory produces about 130 varieties of industrial pumps with 455 specifications. Its physical output rose from 40 pumps in 1952 to 2802 pumps in 1981. The maximum size of an industrial pump is an outlet discharge with a diameter of 64 in. The biggest castings produced by the factory weigh up to 38 tons.

The annual output of the factory is 3000–4000 pumps which are produced mainly on a single shift of 8 hours per day. In the past, the factory used to operate on three shifts. However, under the new economic policy of readjustment, much higher priority is given to agriculture and light industry. According to the Vice Chief Engineer of the Hunan Bureau of Mechanical Industry (and the President of the Hunan Provincial Society for Agricultural Mechanisation), this policy has led to a reduction in the number of shifts. The considerable amount of excess capacity in the factory seems to have been created by the lack of demand for industrial pumps.

The pumps produced by the factory are used by the following user industries: agriculture, irrigation, light and chemical industries, power industries, and transportation. The factory has the following workshops: casting, welding, processing, assembling, repairing, tool and die shop and workshop for making instruments.

Employment and Productivity

In 1952, the factory started with a labour force of 163 which rose to a total of 1799 in 1981. Of the total labour force, there are 39 engineers and 69 technicians. Women represent a sizeable proportion of the total work-force: they are mainly engaged in operating lathes.

As we discussed in Chapter 2, under the Chinese labour system, it is still not possible to lay off any labour. Therefore there may have been more workers in the factory than were technically necessary. The employment of a staff of nearly 2000 in 1981 was reported to be the highest ever since the factory started production.

The average labour productivity in the factory appeared to be quite low. This was explained not so much by the lack of equipment as the poor work organisation, unsatisfactory plant layout, and waste of space in different departments. However, there seemed to be a reasonable amount of workflow. The workers were spared the burden of carrying heavy machinery by the installation of an overhead crane system. Nevertheless the working environment and conditions of workers' safety were rather poor with hardly any workers wearing safety glasses.

The factory buildings were rather old and poorly maintained. But the factory had ceiling fans for use in the summer months when temperatures are quite high.

Technology

The industrial operations of the factory were mainly labour-intensive. Technology has not changed very much for several years. The bulk of the equipment in the factory was of local design. Most of their indigenous machinery was manufactured in Shanghai and Beijing. A few machines imported from Japan, Eastern Germany and the Soviet Union were also being used in the factory. As the factory was established in 1951, it is possible that some of its equipment was built according to Russian blueprints and specifications. The equipment was of 1950–60 vintage with no major replacements since then with the exception of the introduction of Japanese machinery. The Chinese variant of the Japanese lathes were also used side by side with the imported machinery.

The factory had standard designs of its own, but it also manufactured to order according to customers' specifications.

In view of its recent export drive, the factory was increasingly concerned with standardisation and quality control of products. The quality control was ensured by the Hunan Provincial Bureau of Mechanical Industry.

Export Potential

The factory started exporting its products in 1975. So far, its exports have been quite modest. Between 1975 and 1981, only 1700 pumps have been exported to 32 countries around the world. Recently, German, Thai and US companies have visited the factory with a view to examining its products and to placing orders. The Ingersoll–Rand Co. of the USA ordered two modern pumps. We were told that the US company was satisfied with the factory's product quality and had signed a contract for the supply of 50 pumps during 1981. In 1980, the company had bought 60 castings made specially according to US specifications. Machining was done in the USA in view of the special tolerance and specification required.

The factory made contacts with the foreign buyers through the Hunan Provincial Import and Export Corporation which acts as an intermediary and assists the industrial enterprises in producing for exports according to international quality standards. It was not clear whether the factory was entitled to retain the foreign exchange earnings from exports for its expansion and modernisation. It is possible that the factory either shared these earnings with the provincial corporation or that the bulk of them are remitted to a pool controlled by the central government.

Pistons and Piston Rings Factory (Changsha)

The factory was established before the liberation in 1943. Since then it has developed rapidly: so much so that it recently bought an industrial licence from a company in the UK. It covers an area of 2000 square metres and employs a staff of 290 including technical personnel and unskilled labour. It has a total investment of about 17.15 million yuan. The factory owns a stock of 439 pieces of equipment.

The factory is a leading producer of piston rings in the country. It is one of the 80 medium to large factories (of which 25 are large-scale) in China for producing pistons. There are 18 factories for producing piston rings. The majority of piston rings produced by the factory are of high quality and meet the national standards.

Output

The annual output of the factory is 13.6 million piston rings and 690 000 pistons. This output consists of 40 types of pistons and 100 types of piston rings. The factory distributes its output to 92 engine factories in 29 provinces. Most of its output is destined for the domestic market although a portion of it is also exported.

The factory operates at only 50 per cent of its capacity which, we were told, can be raised by about 40 per cent if the market for its product can be assured. The factory operates on three shifts of eight hours a day, six days per week.

Licensing Agreement

The factory exchanges know-how and technical information with Japan and the UK. Since 1979, it has signed a licensing agreement with Wellworthy Ltd, a UK-based firm. In 1980, the factory sent its engineers to the UK for advanced training; a few technicians were also sent for training in 1981. The staff of the UK company visited the factory in 1981 to provide training courses at the factory site and to set up experimental production lines. The factory has modified its plant structure in the light of the recommendations made by the British firm. We were told that the British firm was now fully satisfied with the performance of the factory.

The agreement with the British firm covers the supply of equipment and know-how as follows:

Know-how/technical information $250 000
Equipment (from Britain and Japan) $250 000
Total cost $500 000

Although our Chinese hosts described the agreement as an industrial licence, it seems to be a case mainly of an exchange of know-how under which royalty payments are made. No licence is patented for the use of the British process. Royalty payments are made in cash with a 30 per cent payment during the first year, and the remainder payable over a period of eight years. During the period of the agreement, no interest is charged on credit offered by the British firm. Under the terms of the agreement, the British firm is expected to supply all new technical information that is likely to become available during the eight-year period. These terms and conditions seem to be quite reasonable and favourable to the Chinese factory.

The contractual agreement with the UK firm also covers visits by the Chinese factory staff to the UK and the cost of the training courses. The contract was negotiated by the factory manager and his senior staff in close collaboration with the National Ministry of Machine Building.

The know-how transfer under the terms of the contract included temperature control and the composition of chemical solutions for plating. The Chinese factory decided to approach the British firm only after a Japanese firm had failed to supply the required technology. Prior to signing the contract with the British firm, the Provincial Government of Hunan approached various research institutes throughout China to explore whether any one of them could provide the required chemical solutions.

The factory manager believed that the factory's level of technical know-how was rather low. However, he was confident of making rapid improvements through agreements with the foreign firms like the one with Wellworthy of UK.

The future production plans of the factory are at present based on the forecasts of domestic market alone. However, the factory plans to enter into the export market eventually. For this purpose, it has already signed some contracts with a few Far Eastern countries.

Workers' Welfare

The average monthly wage per employed worker in the factory is 70 yuan. The workers also receive fringe benefits like housing accommodation at a nominal cost of 1 per cent of the wage. Medical care and children's schooling are free.

Spark Plugs Factory (Zhu Zhou)

Established in 1961, this factory produces plugs for farm machinery, cars, torches and so on. It covers an area of 100 000 square metres, and employs 1000 workers. It is one of 427 spark plugs factories in Hunan.

Output

The factory has an annual capacity of 10 million pieces of spark plugs of 33 different varieties and specifications. The plant envisages expansion of production through 'compensation trade' (coproduction with foreign partners). Under this arrangement the plant can obtain capital equipment and intermediate goods from abroad in exchange for exports of its finished products. The products of the factory are sold throughout China and abroad (particularly in Africa and South–East Asia).

The factory normally operates on two shifts per day although some of its workshops also operate on three shifts, especially if the nature of work requires continuous operation.

Technology

The factory manager told us that the level of mechanisation of operations in the factory was rather low, thus necessitating the use of a lot of labour. Currently, efforts are being made to modernise equipment through the importation of technology and know-how from the advanced countries. The factory has already installed very automatic turret lathes which are capable of doing more than seven operations without any manual control. Only skilled monitoring of the machine is required from time to time.

Most of the machinery used in the factory was manufactured in China (in Shanghai and the Northern Province). In the early 1960s, the first few machines of an older vintage were also produced in Hunan at a time when it was impossible to import machinery from other provinces. Subsequently, more modern equipment was imported from the rest of China to modernise factory production, especially with a view to expanding exports.

The porcelain department of the factory at present makes porcelain parts using manually operated machines which seemed to operate quite fast. However, plans are under way to replace these by automatic machines, the prototypes of which have been developed in the R and D department of the factory.

The packaging of spark plugs into cartons is done manually by a large number of boys and girls with tremendous speed. The workers seemed to be quite experienced in their job of packaging and material-handling.

The factory manager explained to us that he was interested in introducing further mechanisation particularly in ceramics manufacture and assembling of parts. The other operations are already well mechanised. The plant management seemed to be aware of the serious unemployment problem in China, and the possible labour-displacement effect of mechanisation. Nevertheless, it felt that additional mechanisation was essential for raising productivity and product quality. An example of recent mechanisation given to us was that of a new machine for welding in order to speed up the production process. It is not clear whether mechanisation is really economical for the factory. For instance, mechanisation of assembling of parts may not necessarily lead to quality improvements or a rise in labour productivity which could just as well be achieved through better work organisation.

R and D Department

The factory has a research and development department which is engaged in testing and experimentation of equipment and prototypes. The management told us that this department suffered from shortages of scientific manpower: in particular, that of technicians. This was one of the few factories that we visited in which proper emphasis was placed on research and application of innovations on the shopfloor. The factory placed stress on the quality control of its products in which the research department plays an important role. All the products are designed and manufactured within the factory, which won a major national award (a silver medal) in 1980 for producing high-quality plugs. In 1981, the factory had surpassed the national standard.

The R and D department plays an important role in improving the quality of the plant's products and processes. Three per cent of the annual budget of the factory goes into R and D. The state allocates additional funds for major research and development projects.

Welding Factory (Zhu Zhou)

Established in 1958, this factory is one of the key factories under the responsibility of the First Machinery Ministry. It has a staff of 1118 workers and 75 technicians. It covers an area of 1400 square metres.

Output

The annual output of the factory is valued at 260 million yuan. The output consists of a variety of welding tools with 157 specifications, and 80 different varieties and 40 specifications of welding machines. The factory accounts for only a small share of the total market since there are a large number of welding factories in China.

Although the factory produced continuous welding machines in the past, their production has now been discontinued, presumably for lack of demand. The factory specialises mainly in welding tools although it also produces welding and cutting machinery.

The factory recently started producing for the export market. Five varieties of welders are exported to 11 countries throughout the world.

R and D

The factory has seven workshops including machine tools, electric shops, and forging and dyeing shops. In addition to these workshops, it has an R and D department. In 1981, the factory started development of new product varieties for the first time. Research was in progress into seven new varieties of welders. Three research projects on welding tools were already completed. The factory's technical staff totalled 75 technicians, of whom 25 were primarily engaged in research. The special welding tools for shipbuilding and river bridges, produced by the factory, had a ready market in China. The factory was also engaged in improving its quality standards to compete in the export market. With this objective, the factory invited the Lloyds Shipping Co. of the UK towards the end of 1981 in order to examine and certify its tools.

Cable and Wire Factory (Xiantan)

This factory was established in 1951. It covers an area of 637 000 square metres of which 122 000 square metres are under buildings of old works and 191 000 square metres are in new works. This is the only factory that we visited in Hunan which paid special attention to worker safety. Two red posters with the following mottos decorated the factory entrance: (i) take care of worker safety, and (ii) remember and respect the party principles.

Output

The annual physical output of the factory is 10 000 tons of copper wires and 13 000 tons of aluminium wires, both estimated to be of the value of 120 million *yuan*. The factory produced the following types of wires and cables: copper and aluminium wires, insulation cables, and copper bars and strips. The decisions to produce two different types of wires were determined at two different levels, namely, the Provincial Government Planning Bureau which lays down output targets, and the overall market requirements.

The wires and cables are produced for both domestic and foreign markets. The factory exported its output to Hong Kong and South-East Asia. The factory manager believed that their output had a comparative advantage in the export market in respect of both price and quality. Yet the factory suffered from keen competition from a large number of other manufacturers of wires and cables. In spite of the unsatisfied domestic demand for wires and cables, the factory manager forecast good export prospects for the factory's products.

The factory suffered from large excess capacity. It is doubtful whether it would be economically viable for a highly capital-intensive factory to import its materials to specialise for export production.

Investment

The fixed investment of the factory was estimated at 40 million yuan. Investment during 1978–80 was 10 million yuan, or 25 per cent of the total. Until 1981, all the profits of the factory were handed over to the state. Since the beginning of 1981, however, the factory was allowed to retain a portion of its profits for capital accumulation. In the absence of profit retention in the past, the factory used to receive capital of about 3 million yuan from the state for modernisation of equipment and other technical improvements.

R and D

The factory has seven workshops and four R and D departments. R and D undertaken by the factory related mainly to technology adaptations and to the development of new products. Three per cent of the factory's annual budget is devoted to R and D expenditure.

Technology

It was a highly capital-intensive factory which required the following high technology from outside China: fissions of solid physics, equipment for testing and experimentation, and measurement and precision instruments. The factory is engaged in experiments on materials and factor substitution. It is working on different kinds of alloys to produce wires and cables more cheaply. In some cases, the factory had already shifted from the use of copper (which is scarce and expensive) to that of aluminium.

Employment and Training

The factory has a staff of 3000 which includes 200 engineers and technicians, and 200 managerial and administrative staff. The average factory worker had secondary school education. Among the engineers and technicians, one-third were college graduates, and two-thirds were vocational school graduates. The factory has a staff college for the training of its personnel. The factory found it particularly difficult to recruit technicians who were in serious shortage.

The factory was not allowed to do its own labour recruitment. The Labour Bureau of the Hunan Provincial Government assigned its technical staff. The factory staff could not be laid off, nor could it be easily transferred either within the province or outside. However, inter-provincial transfers were somewhat easier. Inside the plant, labour shifts between different departments were possible, and often encouraged to give greater experience to the factory staff.

The average wage per factory worker is 90 yuan per month. The factory management was unable to provide us with any data on wages by skill categories.

Diesel Engine Factory (Xiantan)

Established in 1965, this factory initially produced old diesel engines. During the 17 years since its inception the factory has produced 107 000 diesel engines of which a certain proportion was exported to Thailand and Hong Kong. The factory has eight workshops, 16 departments/divisions and nine production lines.

Output

The present output of model 175 engine (a single cylinder, four-stroke, water-cooling, horizontal type engine with a rated output of 5 h.p. at 2000 r.p.m.) is 10 000 engines per year. The factory expected to start production of two new models (nos. 185 and RHF 175) which were recently developed in the factory.

Twenty per cent of the factory's total output was exported to Pakistan and Thailand. The factory management believed that their products were of superior quality and thus could successfully compete in the international market.

There is a great demand in China for 2–3 h.p. engines which are ideally suited for the small farmers' needs. The factory planned to produce these lighter engines in the foreseeable future.

The factory's productive capacity is about 20 000 engines per year. Since the actual output per year is well below this figure, the factory seemed to be operating at only half of its rated capacity. The low capacity is partly explained by the old product: the new models have not yet gone into production.

In 1979, the factory was able to sell 20 000 engines (that is, twice its present annual output). However, it could no longer reach that old level since four similar factories now shared an already saturated market for diesel engines.

The factory is making bicycle brakes in order to utilise its excess capacity. It hoped to shift back to the production of diesel engines when the new models were commercialised. We were told that the factory had already received a sufficient number of orders from within and outside China.

Investment

Out of a sales value of 8.51 million yuan, the factory's profits were of the order of 600 000 yuan. The factory is allowed to retain 30 per cent of the total profits for reinvestment and provision of social welfare facilities. Of the retained profits, 30 per cent are reinvested whereas 70 per cent are spent on social welfare.

The cost of the 175 model diesel engine is 460 yuan and its selling price (retail) is 600 yuan.

R and D

The R and D department of the factory is assisted by the Research Institute in Shanghai in the development of new products. The Shanghai institute provided technical assistance and undertook testing and experimentation of the prototypes on behalf of the factory, which has a unique contractual agreement with the institute. The latter cannot provide similar technical assistance to other factories.

The technical staff of the factory engaged in R and D constitutes 5 per cent of the total work-force of 1300 in the factory. There are 20 engineers and 42 technicians.

Two per cent of the factory's annual budget is devoted to R and D for product development.

Technology

The handling and transportation of engines in the factory is quite labour-intensive. Notwithstanding a short conveyor belt installed in the factory recently, the assembled engines are moved into the storage room manually by workers carrying stretcher-trolleys.

The diesel engine produced by the factory is light, compact and easy to maintain. It is suitable for pumping water and for processing farm products. It is also used as a motor for walking tractors, small boats and vehicles. The new models to be produced by the factory have the following characteristics: low weight (about 50 kg), this weight to be reduced by substituting aluminium for cast iron; greater horsepower; air-cooling instead of water-cooling; lower fuel consumption, and better overall performance.

APPENDIX III
HUNAN'S TEXTILE INDUSTRY[1]

During the past 30 years, Hunan's textile industry has raised its output at an annual rate of 13.2 per cent. At present, there are about 320 factories in the province, which are engaged in cotton and ramie weaving, wool spinning, chemical fibre production, knitwear production, dyeing and printing, silk production, and textile manufacturing machinery.

The industry absorbs a total of more than 130 000 workers and staff members, a labour force which is 11 times that in 1949. In 1980, the total output of cotton was 1.96 million which is close to the bumper crop for 1979. The total output of ramie and ambary hemp was more than 1 million *dan*. The Hunan province produces 40 per cent of the national output of ramie (fine strong Chinese fibre). In 1980, it reaped 13 500 tons of the plant from 13 000 hectares, representing an increase of more than 10 per cent over 1979. According to a long-term plan of the Chinese Ministry of Textile Industry, natural fibres like ramie should be fully utilised side by side with the development of chemical fibres. At present, several new mills are being built and the old ones modernised to raise the output of ramie fabrics.[2]

In order to provide an economic incentive for the commune members to raise cotton production, in 1980 the state raised the cotton procurement price by 10 per cent: it is proposed to raise this procurement price by 30 per cent after the fulfilment of the cotton quota. The Hunan Provincial Government organised a Cotton Conference in 1980 to stimulate cotton production and mobilise people in the cotton-growing areas to produce a bumper crop. (Cotton production in 1979 and 1980 failed to meet the total national requirements.)

The textile industry of Hunan in 1979 produced 30 million square metres of dyed cloth. There are 2200 categories of textile goods with more than 13 000 different patterns and specifications.

The Hunan Provincial Bureau of Textile Industry, in consultation with prefectures and municipal departments in charge of local industry, has converted 12 plants of machinery, transport, capital construction and other industries for the production of textiles. This conversion is intended to overcome shortages of textiles and garments. It is reported that seven of the converted plants have already started producing 15 types of textile products including knitwear, underwear and outer garments. In addition, large quantities of raw materials and machine parts and accessories for the industry are also being produced. For example, the CHANGDE prefecture diesel engine plant has been renamed the No. 2 Textile Machinery Plant of the prefecture. Capital investment of 200 000 yuan was made to introduce modernisation and adaptations in equipment and technology. By the end of March 1981, the plant had produced 234 pieces of power-operated knitting machines and 104 looms.[3]

Additional equipment of the order of 63 000 spindles for cotton and 500 looms were provided for the textile industry in 1979. Capital construction in the light and textile industry in Hunan overfulfilled the capital construction investment by 10.5 per cent.[4] In 1980, key capital construction projects included an extension of the Xiantan textile dyeing and finishing mill.

Notes

1. During our visit to Hunan in November 1981, we were unable to visit any textile factories. Data on this industry are therefore not based on field visits. Instead, information was collected from secondary sources.
2. *Xinhua News Agency*, 2 February 1981, and *Changsha, Hunan Provincial Service*, 16 June 1981.
3. *Changsha, Hunan Provincial Service*, 16 April 1981.
4. *Changsha, Hunan Provincial Service*, 20 January 1980.

APPENDIX IV
FARM MACHINERY RESEARCH INSTITUTE (CHANGSHA)

This institute, which is one of several research institutes in the country, was established in 1958. At the time of its establishment, it had a staff of only 50 people, with 17 research engineers and technicians, 12 administrative personnel and 21 workers. In 1981, it had a total staff of 296 divided as in Table A.11 below.

On average, the research staff of the Institute has 4–5 years of education in Beijing University or in other universities and colleges in the country. The director of the Institute was trained in the USSR.

TABLE A.11 *Staff of the Institute*

Category	*Total*	*Men*	*Women*
Engineers and technicians	138	101	37
Administrative personnel	30	106	
Workers	128		52
	296	207	89*

* Including administrative personnel.

SOURCE Data provided by the Institute.

FUNCTIONS

The Institute has six research divisions as follows:

(i) *First Research Division*, doing research on farm power including new energy sources;
(ii) *Second Division*, on tillage, transplanting machinery and equipment for crop protection;
(iii) *Third Division*, on machinery for harvesting and processing agricultural by-products, and equipment for livestock husbandry;
(iv) *Fourth Division*, on equipment for irrigation and drainage;
(v) *Fifth Division*, on measuring techniques for farm machinery and implements;
(vi) *Sixth Division*, on technical information, standardisation, making blueprints and reference materials.

There is also one workshop which manufactures prototype machines and a small amount of farm equipment.

PRIORITIES

This is the only institute of its kind in the whole province, which it has to serve alone. The following are its present research priorities:

(i) 3–5 h.p. pumps of good quality for export to South-East Asia;
(ii) research on sprinkler irrigation which received a top national award;
(iii) small hydro-electric generators;
(iv) small 'walking tractors' which are exported to Thailand.

The priorities for research are determined by the Provincial Bureau for Agricultural Machinery in consultation with the Institute staff. Recently, a national committee of experts has been established to study the manufacture of 3.5 h.p. pumps needed by the small-scale peasants. Staff of different provincial research institutes is represented on it.

A number of prototypes have been developed by the Institute. They are not yet in commercial production, and are unlikely to be produced in the foreseeable future. These are, for example, combine harvesters and threshers which are likely to be de-emphasized under new economic policies which give lower priority to 'heavy industry' than agriculture and light industry. Relatively greater emphasis is placed on smaller and lighter machinery. On display also were the small 'boat tractors' made out of a boat, and run by a small 5 h.p. diesel engine. These tractors were the peasants' innovation which was further researched at the Institute before its commercial production.

The Bureau of Agricultural Machinery has to approve prototypes before they can be mass produced. The performance of different prototypes is controlled by a small Chinese-made 7–K computer. When we visited the Institute, it was in the process of controlling the performance capacity of irrigation pumps.

At present, the farmers in the province are too poor to afford any large agricultural machinery. The experience of rental of tractors and other farm machinery through communes has not been very successful, partly because of lack of access by a large number of peasants. The rental scheme could benefit only a limited number. In the light of these difficulties, the present priority is rightly placed on the development of very small (2 h.p.) internal combustion diesel engines which the small peasants could afford.

LINKAGES

Within the country, the Hunan Institute has established links with the Beijing Institute of Agricultural Machinery and with other provincial research institutes. There is a healthy competition between provincial institutes which motivates researchers to improve their performance. This competition may also have a negative effect in the sense that each provincial institute may try to promote its own models throughout the province whether they are economical or not. If they were less economical than a model developed in another province, it would be more economical to use that model instead. It would be the task of the Central Ministry of Farm Machinery to standardise models and promote the production of only those that are most efficient and cost the least. However, it is not clear whether under the new system of decentralisation and delegation of responsibility to the provinces, such a central intervention would still be encouraged. As the system in general seems to be also fairly centralised, there is nothing to stop the Central Ministry from reducing the large variety of different models produced by provincial institutes in order to promote more specialised production.

Outside China, links have been established with institutions in Japan, and with the International Rice Research Institute (IRRI). The Director of the Institute went on a study tour to Japan.

During the past 22 years, the Institute has completed 226 research projects. Many of the farm machines and implements, such as water pumps, small engines and two-wheel tractors, were developed by the Institute in cooperation with factories and research institutes subordinate to prefectures and counties.

Notes and References

1 THE PROVINCIAL ECONOMY

1. 'Diversify the Rural Economy (Hunan Province)', *Beijing Review*, 7 September 1979.
2. *Changsha, Hunan Provincial Service*, 25 December 1980.
3. *Changsha, Hunan Provincial Service*, 17 June 1982.
4. *Wuhan Hubei Provincial Service*, 30 December 1981.
5. *Changsha, Hunan Provincial Service*, 24 November 1981.
6. *Changsha, Hunan Provincial Service*, 20 May 1980.
7. *Xinhua News Agency*, 30 September 1982.
8. *Changsha, Hunan Provincial Service*, 12 July 1981.
9. *Changsha, Hunan Provincial Service*, 5 November 1981.
10. 1 catty = 0.5 kg.
11. *Changsha, Hunan Provincial Service*, 5 November 1981.
12. 1 picul = 1 dan = 50 kg.
13. *Changsha, Hunan Provincial Service*, 5 June 1981.
14. Interview cited in *Changsha, Hunan Provincial Service*, 11 June 1981. Translated in *SWB*, FE/W 1139/A/4, 24 June 1981.
15. Shigeru Ishikawa, 'China's Food and Agriculture: Performance and Prospects', in Erwin M. Reisch (ed.), *Agriculture Sinica* (Berlin: Duncker & Humbolt, 1982).
16. I-Chuan Wu, 'Nourrir Le Peuple: Politique Alimentaire et Politique Démographique en Chine', *Revue Tiers-Monde*, vol. XXII (April–June 1981) no. 86, table 4.
17. Shigeru Ishikawa, 'China's Food and Agriculture'.
18. Wong Yongxi *et al*, 'Views on Strategic Problems in China's Agricultural Development', *JJYJ*, November 1981.
19. *Changsha, Hunan Provincial Service*, 8 July 1981. Electricity generation in the province is reported to have saved nearly 30000 tons of coal during the first six months in 1981.
20. *Changsha, Hunan Provincial Service*, 23 October 1981.
21. *Changsha, Hunan Provincial Service*, 24 November 1981.
22. *NCNA*, 4 February 1982.
23. *Xinhua News*, Beijing, 4 February 1981.
24. *Changsha, Hunan Provincial Service*, 24 October 1981.
25. *NCNA*, 10 January 1982.
26. *NCNA*, 10 January 1982.
27. *Changsha, Hunan Provincial Service*, 23 July 1981.

28. Liang Liguang, 'Develop Regional Superiority, Greatly Develop Light Industry–A Survey of the Light Industries in Hubei and Hunan', *Congren Ribao* (*Workers' Daily*), Beijing, 17 July 1980.
29. The Chengfeng Production Team of the Huanhua commune in the suburbs of Changsha has 62 families and 259 members. Of these over 20 families had savings of more than 500 yuan in the credit cooperatives in 1979. Twenty-six families had bought 26 bicycles, 30 families had bought 56 watches, nine families had bought nine sewing machines, and 21 families had bought 21 radios.
30. Andrew Heyden, 'The Modern Ceramics Trade', *The China Business Review* (September–October 1982).
31. *Beijing Review*, 21 April 1979, p. 11.
32. Richard Baum (ed.), *China's Four Modernisations: The New Technological Revolution* (Boulder, Colorado: Westview Press, 1980).
33. Dinyar Lalkaka, *Urban Housing in China* (Beijing, June 1983, mimeo.), and A. S. Bhalla and G. A. Edmonds, 'Construction Growth and Employment in Developing Countries', *Habitat International*, Oxford, vol. 7, nos 5–6, 1983.
34. *Changsha, Hunan Provincial Service*, 1 June 1980.
35. Hsiang Jung and Chin Chi-chu, 'Not for Profits: Socialist Commerce (III)', *Peking Review* (30 July 1976) no. 31. In Hengyang in Hunan, which is a medium-sized city with about half a million population, numerous examples of such cooperation have been noted by these two authors.
36. Hsiang Jung and Chin Chi-chu, 'A Vast Rural Market: Socialist Commerce (IV)', *Peking Review* (9 August 1976) nos 32-3.
37. See Uma Lele, 'Rural Marketing in China: A Comparative Perspective', *World Development*, vol. 6 (May 1978) no. 5, and Udo Weiss, 'China's Rural Marketing Structure', *World Development*, vol. 6 (May 1978) no. 5.
38. *NCNA*, 11 March 1979, quoted in Bill Brugger, 'Rural Policy', in Bill Brugger (ed.), *China Since the Gang of Four* (London: Croom Helm, 1980) p. 165.
39. Private catering services are also on the increase. In Xin-xiang, a major city in Henan province, the private share in catering has risen to 48 per cent with 44 per cent for collectives and 8 per cent for the state. In the municipalities of Guilin (Guangxi), Shanchiu (Henan) and Liuan (Anhui), private business makes up two-thirds of catering shops. See Liang Chuanyuan, 'Problems Concerning the Recovery and Development of Individual Industrialists and Businessmen', *JJGL* (1980) no. 7.
40. *Changsha, Hunan Provincial Service*, 1 June 1980.
41. Thomas B. Wiens, 'Price Adjustment, the Responsibility System and Agricultural Productivity', *American Economic Review*, Papers and Proceedings (May 1983).
42. *Changsha Hunan Radio*, 20 April 1979. Quoted in *SWB*, FE/6101/B 11/4–5.
43. Bill Brugger, 'Rural Policy', pp. 163–7.
44. Liang Wensen, 'Balanced Development of Industry and Agriculture', in Xu Dixin *et al.* (eds), *China's Search for Economic Growth – The Chinese Economy since 1949* (Beijing: New World Press, 1982) p. 71.

45. Carl Riskin, *The Terms of Trade Between Industry and Agriculture in China* (1983, mimeo.).
46. Liang Wensen, 'Balanced Development of Industry and Agriculture'.
47. Thomas B. Wiens, 'Price Adjustment'.

2 THE NEW ECONOMIC POLICY FRAME

1. For a general discussion of the recent Chinese economic reforms, see Xue Muqiao, *China's Socialist Economy* (Beijing: Foreign Language Press, 1981); Xue Muqiao, 'Tentative Study on the Reform of the Economic System', *Chinese Economic Studies*, New York, vol. XIV (Winter–Spring 1980–1) nos. 2–3; Sung Tao, 'On China's New Economic Policies', *China Enterprise*, Hong Kong, December 1981; He Jianzhang, 'Newly Emerging Economic Forms', *Beijing Review*, 25 May 1981; and Xia Zhen, 'A New Strategy for Economic Development', *Beijing Review*, 10 August 1981.
2. See Robert F. Dernberger, 'The Chinese Search for the Path of Self-sustained Growth in the 1980's: An Assessment', in US Congress Joint Economic Committee, *China Under the Four Modernisations*, Part I (Washington, D.C., August 1982).
3. See Yang Jianbai and Li Xuezeng, 'The Relations Between Agriculture, Light Industry and Heavy Industry in China', *Social Sciences in China*, vol. I (June 1980) no. 2.
4. Wang Bingqian, 'Report on Financial Work', *Beijing Review*, 29 September 1980.
5. See *Beijing Review*, 25 October 1982, and Ajit Kumar Ghose, *The New Development Strategy and Rural Reforms in Post-Mao China* WEP Working Paper, WEP 10-6/WP. 62 (ILO, Geneva, November 1983).
6. *Changsha, Hunan Provincial Service*, 2 December 1981.
7. Zhang Shugyuang, 'Reform the Economic Structure and Increase the Macroeconomic Results – An Analysis of China's Economic Structure and the Changes in Economic Results', *Social Sciences in China*, Beijing, vol. III (March 1982) no. 1.
8. Ibid.
9. 9 May 1981. Speech by Xue Muqiao at the First Sino-Japanese Meeting for the Exchange of Economic Knowledge, *Shanghai 'Shijie Jingji Daobao'*, (1 June 1981) no. 35; translated in *SWB*, FE/W 1140/C/1, 1 July 1981.
10. Agricultural sideline activities and light industry may often compete for the same raw materials (see Chapter 6).
11. In order to correct the effects of arbitrary differences in profitability between industries and enterprises, the Chinese policy-makers vary profit retention 'ratios' so that industries/enterprises with low prices and profits are allowed to keep a high share of profits, whereas those with high profits would be allowed a lower retention rate. (See Carl Riskin, 'Market, Maoism and Economic Reform in China', in Victor Lippit and Mark Selden (eds), *The Transition to Socialism in China* (White Plains, N.Y.: M.E. Sharpe, 1982).)

12. Joint Operations in Light Industry, *Peking Home Service*, 23 May 1980. *SWB*, FE/W 1085/A/10, 4 June 1980.
13. Andrew Watson, 'Industrial Development and the Four Modernisations', in Bill Brugger (ed.), *China Since the Gang of Four* (London: Croom Helm, 1980) pp. 118–19.
14. See Liang Liguang, 'Develop Regional Superiority, Greatly Develop Light Industry – A Survey of the Light Industries in Hubei and Hunan', *Congren Ribao* (*Workers' Daily*), Beijing, 17 July 1980.
15. Quoted in *International Herald Tribune*, 28–9 November 1981.
16. On average, the commune enterprises employ over 40–50 workers and cover such activities as consumer goods industries, machine shops, cement plants, tractor-servicing stations, and so on. A commune consists of about 12–13 brigades. The brigades, which usually consist of 7–12 teams, carry out economic and social functions similar to those of the communes, but on a smaller scale.
17. *Xinhua News*, 24 January 1983.
18. See Amartya Sen, *Employment, Technology and Development* (Oxford: Clarendon Press, 1975) chapter 3.
19. See Alexander Eckstein, *China's Economic Revolution* (Cambridge University Press, 1977) pp. 82–3.
20. See J. Sigurdson, *Rural Industrialisation in China* (Cambridge, Mass.: Harvard University Press, 1977); C. Riskin, 'Intermediate Technology in China's Rural Industries', in Austin Robinson (ed.), *Appropriate Technology for Third World Development* – Proceedings of a Conference held by the International Economic Association (London: Macmillan, 1979); and Keith Griffin and Ashwani Saith, *Growth and Equality in Rural China* (Singapore: ILO, Asian Employment Programme (ARTEP), 1981).
21. See Shigeru Ishikawa, 'China's Food and Agriculture: Performance and Prospects', in Erwin M. Reisch (ed.), *Agriculture Sinica* (Berlin: Duncker & Humboldt, 1982).
22. Thomas B. Wiens, 'Price Adjustment, the Responsibility System and Agricultural Productivity', *American Economic Review, Papers and Proceedings* (May 1983).
23. Wang Shoudao, 'The Responsibility Systems have brought New Vitality to the Rural Areas of Hunan – Report of An Investigation of Several Prefectures and Counties in Hunan', *RMRB*, Beijing, 5 June 1981.
24. Katsuhiko Hama, 'China's Agricultural Production Responsibility System', *China Newsletter*, JETRO, Tokyo, (September–October 1982) no. 40. Also see Graham E. Johnson, 'The Production Responsibility System in Chinese Agriculture: Some Examples from Guangdong', *Pacific Affairs*, vol. 55 (Fall 1982) no. 3; Jürgen Domes, 'New Policies in the Communes: Notes on Rural Societal Structures in China, 1976–1981', *Journal of Asian Studies*, vol. XII (February 1982) no. 2; and Azizur Rahman Khan, 'The Responsibility System and the Change in Rural Institutions', in Azizur Rahman Khan and Eddy Lee, *Agrarian Policies and Institutions in China after Mao*, ILO/Asian Employment Programme (Maruzen, Singapore, 1983).

25. These two types of the responsibility system are likely to be particularly suitable for relatively poorer villages and production teams in which widely scattered households could still carry on production on a small scale.
26. *NCNA*, 23 March 1981.
27. *International Herald Tribune*, 14 April 1983.
28. Jürgen Domes, 'New Policies in the Communes'.
29. Katsuhiko Hama, 'China's Agricultural Production Responsibility System'.
30. See Carl Riskin, 'Market, Maoism and Economic Reform in China'.
31. For a detailed discussion of the system, see Akira Fujimoto, 'The Economic Responsibility System in China's Industrial Sector', *China Newsletter* Tokyo: JETRO (November–December 1982) no. 41.
32. 'Beijing's Pilot Project in Commerce – Events and Trends', *Beijing Review*, 31 January 1983.
33. During the three years between 1979 and 1981, 32.9 per cent of those employed worked in collective enterprises, and 3.8 per cent were self-employed.
34. Zeng Qixian, 'Employment Creation and Economic Development', in Xu Dixin *et al.*, *China's Search for Economic Growth — The Chinese Economy since 1949*, China Studies Series (Beijing: New World Press, 1982).
35. Benedict Stavis, 'Rural Institutions in China', in R. Barker, R. Sinha *et al.* (eds), *The Chinese Agricultural Economy*, Westview Special Studies on China and East Asia (Boulder, Colorado: Westview Press, 1982).
36. *Beijing Review*, 23 November 1979.
37. The rustication movement is known to continue although at a more moderate pace than in the Mao era. For a more detailed analysis of this movement, see Leo A. Orleans, 'China's Urban Population: Concepts, Conglomerations and Concerns', in US Congress Joint Economic Committee *China Under the Four Modernisations*.
38. *International Herald Tribune*, 14 April 1983.
39. See Hu Mengzhou, 'Solution to Employment Problems', *Beijing Review*, 27 September 1982, and Feng Lanrui and Zhao Lükuan, 'Urban Unemployment in China', *Social Sciences in China*, vol. III (March 1982) no. 1.
40. *Changsha, Hunan Provincial Service*, 23 September 1981.
41. *Changsha, Hunan Provincial Service,* 30 January 1983.
42. *Changsha, Hunan Provincial Service*, 24 January 1982.
43. Wu Jiuxin, 'Beijing is Solving its Youth Unemployment Problem', *China Reconstructs*, vol. XXXII (June 1983) no. 6, and 'Reform of the Employment System', *Beijing Review*, 4 April 1983.

3 THE TECHNOLOGY POLICY FRAME

1. Chu Xiangyin and Wang Shouchun, 'The Problem of Developing Our Export of Labour-intensive Products', *International Trade Journal* (*Guoji Maoyi Wenti*), (Winter 1981) no. 4.

2. Alexander Eckstein, *China's Economic Revolution* (Cambridge University Press, 1977) pp. 152–8.
3. Shigeru Ishikawa, 'A Note on the Choice of Technology in China', *Journal of Development Studies*, vol. 9 (October 1972) no. 1.
4. Wang Enkui, 'A Discussion of Technological Economics (*Tantan jishu jingji*)', *HQ* (1980) no. 7, pp. 11–13. Also see Erik Baark, 'China's Technological Economics', *Asian Survey*, September 1981.
5. Maurice Dobb, *An Essay on Economic Growth and Planning* (London and New York: Modern Reader, 1960).
6. Shigeru Ishikawa, 'A Note on the Choice of Technology in China'.
7. The period 1961–5 which referred mainly to readjustment and recovery in the wake of the Great Leap is ignored here.
8. See William W. Hollister, 'Trends in Capital Formation in Communist China', in *An Economic Profile of Mainland China*, vol. I, Studies prepared for the Joint Economic Committee, Congress of the United States (Washington, D.C., February 1967).
9. The data for fixed assets per worker for different state and state–private enterprises for 1952–3 are as follows: 51 197 yuan for electric power followed by 24 945 yuan for petroleum, 9528 yuan for paper, 5029 yuan for coal and 4806 yuan for textiles compared to the figure of 5656 yuan as average for all industry. See Thomas G. Rawski, *China's Transition to Industrialism; Producer Goods and Economic Development in the Twentieth Century* (Ann Arbor, Michigan: University of Michigan Press, 1980).
10. Thomas G. Rawski, 'Choice of Technology and Technological Innovation in China's Economic Development', in Robert F. Dernberger (ed.), *China's Development Experience in Comparative Perspective* (Cambridge, Mass.: Harvard University Press, 1980).
11. Genevieve C. Dean, *Technology Policy and Industrialisation in the People's Republic of China,* IDRC. STPI, no. 4 (Ottawa, 1979), p. 39.
12. In his speech 'On the Ten Major Relationships' in 1956, Mao criticised the undue importance given to heavy industry and the relative neglect of agriculture and light industry.
13. See *Xinhua News Agency*, Daily Bulletin, 14 November 1958; and *SCMP*, 21 August 1960, 9 November 1961, and 25 December 1961, quoted in E.L. Wheelwright and Bruce McFarlane, *The Chinese Road to Socialism* (Harmondsworth: Penguin, 1970).
14. E.L. Wheelwright and Bruce McFarlane, *The Chinese Road to Socialism*, p. 173.
15. See Yang Jianbai and Li Xuezeng, 'The Relations Between Agriculture, Light Industry and Heavy Industry in China', *Social Sciences in China*, vol. I (June 1980) no. 2, p. 200.
16. Thomas G. Rawski, *Economic Growth and Employment in China* (Oxford University Press, 1979) pp. 48–9.
17. See World Bank, *World Development Report for 1982* (Washington, D.C.: Oxford University Press and the World Bank, 1982) p. 116.
18. 'We Must Pay Particular Attention to Consumption Growth', in *Jingjixue Dongtai* (*Trends in Economic Studies*), Beijing, S Wen Zhai Bao, 22 June 1982. (Translated in *FBIS*, Daily Report on China, *FBIS*-CHI–82–135, 14 July 1982, vol. 1, no. 135.)

19. For a general discussion of science policy and industrial research in historical perspective, see OECD, *Science and Technology in the People's Republic of China* (Paris, 1977).
20. For a discussion of vertical administrative management, see Erik Baark, *Techno-economics and Politics of Modernisation in China: Basic Concepts of Technology Policy under the Readjustment of the Chinese Economy*, Discussion Paper no. 135 (Research Policy Institute, Lund University, November 1980) p. 32; and Liao Jili, 'A Discussion of Organisational Change from the Viewpoint of Horizontal and Vertical Relations', *RMRB*, 26 August 1980, p. 5.
21. See Zhou Jizhong and Chen Cangye, 'We Should Establish Development Research Centres', *Guangming Ribao*, 1 August 1980.
22. See Daniel Chudnovsky and Masafumi Nagao, *Capital Goods Production in the Third World – An Economic Study of Technology Acquisition* (London. Frances Pinter, 1983), chapter on China.
23. *Changsha, Hunan Provincial Service*, 19 May 1981.
24. For a discussion of this issue, see A. S. Bhalla, 'Internal Technology Transfers in China', paper presented at the International Conference on Science and Technology Policy and Research Management, organised by the Chinese State Science and Technology Commission in conjunction with the UN Financing System on Science and Technology for Development (Beijing, 4–8 October 1983).
25. Daniel Chudnovsky and Masafumi Nagao, *Capital Goods Production in the Third World*.
26. Jon Sigurdson and Boel Billgren, 'An Estimate of Research and Development Expenditure in the People's Republic of China in 1973', *Industry and Technology Occasional Paper no. 16* (Paris: OECD Development Centre, July 1977) p. 7.
27. Richard P. Suttmeier, *Science, Technology and China's Drive for Modernisation* (Stanford University, California: Hoover International Studies, 1980) pp. 62–6.
28. These figures are taken from Jon Sigurdson, *Technology and Science in the People's Republic of China – An Introduction* (Oxford: Pergamon, 1981).
29. It should be noted that promotions and salary increases were frozen for scientific and professional personnel for more than ten years during the Cultural Revolution. See Richard P. Suttmeier, *Science, Technology and China's Drive for Modernisation*.
30. *Current Literature on Science of Sciences*, National Institute for Science, Technology and Development Studies, New Delhi, vol. 11 (March 1982) no. 3, p. 58.
31. 'Foreigners are Dangerous – Official', *Far Eastern Economic Review*, 1–7 October 1982.
32. *Xinhua News Agency*, 15 January 1983.
33. *NCNA*, 3 May 1980.
34. *Changsha, Hunan Provincial Service*, 2 April 1981.
35. Erik Baark, *Techno-economics and Politics of Modernisation*, pp. 22–3.
36. *NCNA*, 21 July 1980; cited in *SWB* FE/W1104/A/11, 15 October 1980.
37. See John N. Hawkins, 'Rural Education and Technique Transformation

in the People's Republic of China', *Technological Forecasting and Social Change* (1978) no. 2.

38. During his visit to the Hua-Tung commune, Ward Morehouse also found 'no organisational structure for technological innovation in the rural industrial sector' parallel to that for agricultural research. See Ward Morehouse, 'Notes on Hua-Tung Commune', *China Quarterly* (September 1976) no. 67.
39. See Jon Sigurdson, *Rural Industrialisation in China* (Cambridge, Mass.: University Press, 1977) chapter 3.
40. Nevertheless there is some horizontal transfer of technology among commune and county enterprises, see ibid., pp. 89–90.
41. *Modernisation of Industry Related to Agriculture in Hunan (China),* Papers and Proceedings of the Changsha International Seminar (Changsha, 4–14 November 1981), Volunteers in Technical Assistance (VITA) (Arlington, Virginia, USA, April 1983).
42. Denis Fred Simon, 'China's Capacity to Assimilate Foreign Technology – An Assessment', in US Congress Joint Economic Committee, *China Under the Four Modernisations*, Part I (Washington, D.C., 1982) p. 547.
43. *Joint Publications Research Service* (JPRS), 78044 (1 May 1981) pp. 1–4.

4 TECHNOLOGY IMPORTS AND FOREIGN INVESTMENT

1. Robert Michael Field, 'Real Capital Formation in the People's Republic of China, 1952–1973', April 1966, p. 60 (unpublished manuscript).
2. Kang Chao, *The Construction Industry in Communist China* (Edinburgh University Press, 1968) p. 61.
3. Thomas G. Rawski, 'Choice of Technology and Technological Innovation in China's Economic Development', in Robert F. Dernberger (ed.), *China's Development Experience in Comparative Perspective* (Cambridge, Mass.: Harvard University Press, 1980).
4. Chen Weiqin, 'The Direction of Introducing Technology should be Changed', *JJGL* (15 April 1981) no. 4.
5. United Nations, *Trade Statistics, 1970–1979* (New York, 1980).
6. See Ren Jianxin, 'Legal Aspects of China's Technology Import and Utilisation of Foreign Investment', *Economic Reporter*, Hong Kong (September 1980) no. 9.
7. *SCR* 1980 (20 November 1980) no. 14, p. 419; cited in Bruce L. Reynolds, 'Reform in Chinese Industrial Management: An Empirical Report' in US Congress Joint Economic Committee, *China Under the Four Modernisations*.
8. *China Economic News*, 29 March 1982.
9. Christopher M. Clarke, *China's Provinces: An Organisational and Statistical Guide* (Washington, D.C.: National Council for US–China Trade, 1982) p. 192.
10. *Changsha, Hunan Provincial Service*, 23 February 1981.
11. *China Economic News*, 22 March 1982.
12. *Xinhua News Agency*, 22 October 1982.

13. *Xinhua News Agency*, 15 January 1983.
14. 'China's High Ranking Economic Official on the Chinese Investment Promotion Meeting', *Economic Reporter* (April 1982) no. 4.
15. See A. S. Bhalla, 'Technological Choice in Construction in Two Asian Countries: China and India', *World Development*, March 1974.
16. Teng Weizao, 'Socialist Modernisation and the Pattern of Foreign Trade', in Xu Dixin *et al.* (eds), *China's Search for Economic Growth – The Chinese Economy since 1949*, China Studies Series (Beijing: New World Press, 1982) p. 181.
17. According to Chen Weiqin, the average price of the steel rolling machines imported by the Baoshan iron and steel plant from Western Germany is 16000 dollars per ton, not including 12 to 20 per cent of monopoly fees and other fees for technical materials and designs. In contrast to this, the price per ton of a domestic steel rolling machine is only 5–6000 yuan (see Chen Weiqin, 'The Direction of Introducing Technology should be Changed').
18. In other words, application of the concept of a 'payout' or 'recoupment period' discussed earlier. However, although the concept of a rate of return on investment is inconsistent with the Marxist theory under which capital is simply embodied labour and not an independent factor of production, recent Chinese writings suggest that this concept is now being considered, at least implicitly, in decisions on investment allocation and technical choice.
19. *Changha, Hunan Provincial Service*, 25–7 December 1980.
20. *China Economic News*, 22 March 1982.

5 AGRICULTURAL MECHANISATION

1. Rakhal Datta, 'Technology Choice in Collectivised Agriculture: Farm Mechanisation Policy of the People's Republic of China', Special Articles, *China Report*, New Delhi, vol. XVI (September–October 1980) no. 5.
2. Zhai Yao, 'Problems in the Mechanisation of China's Agriculture', *Economic Reporter*, Hong Kong (July 1981) no. 7.
3. Dwight Perkins *et al, Rural Small-Scale Industry in the People's Republic of China*, The American Rural Small-scale Industry Delegation (University of California Press, 1977) p. 118.
4. *RMRB*, 8 October 1981, p. 2. Also see Robert C. Hsu, *Food For One Billion–China's Agriculture since 1949* (Boulder, Colorado: Westview Press, 1982) p. 130.
5. While this is possible in manufacturing subject to certain limitations, our visit to the Hunan Agricultural Machinery Research Institute at Changsha indicated that a large number of prototypes of large tractors and combine harvesters would now remain at the laboratory stage with little chance of their commercial production. This would result in a considerable waste of scarce capital.
6. Large-scale importation of heavy agricultural machinery into China for the past several years may have discouraged the development of indigenous small-scale farm machinery industry. In the 1950s, the bulk of

machinery imports came from the Soviet Union. Since 1978, China has introduced 17 whole sets of seed-to-harvest equipment and more than 1300 individual machines that were imported from the Federal Republic of Germany, Japan, United States and other countries for use on dry crops, paddy rice, industrial crops like cotton and rubber, and animal farming (see Zhai Yao, 'Problems in the Mechanisation of China's Agriculture').

7. *NCNA*, 11 December 1981. Quoted in *SWB*, Second Series, FE/W 1166/A/36, January 1982.
8. 'It is Essential to Attach Importance to Farm Machinery Management', an article in *People's Daily*, 18 March 1980. Translated in *SWB*, FE/W 1079/A/10, 23 April 1980.
9. World Bank, *China: Socialist Economic Development*, vol. II, Washington, D.C., August 1983, pp. 74–5.
10. Statement by Yang Ligong, Chinese Minister of Agricultural Machinery, in an interview with Peking Radio, *Peking Home Service*, 9 March 1980. Quoted in *SWB*, FE/W 1074/A/3, 19 March 1980.
11. *NCNA* (in Chinese), 13 January 1980. Quoted in *SWB*, FE/W 1066/A/16, 23 January 1980.
12. Xia Deji, 'Agricultural Mechanisation in Hunan', paper presented at the International Seminar on the Modernisation of Industry Related to Agriculture, Changsha, Hunan, 4–14 November 1981.
13. The equipment data in the table does not seem to include the output of small–scale agricultural machinery industry in the rural areas. For example, the American Rural Small-scale Industry Delegation (1975) noted that they were unable to discuss the rural-based farm equipment industry, and that their comments on agricultural mechanisation were based primarily on 'the agricultural machines ... being produced in the manufacturing plants and used in the progressive agricultural areas'. (See Dwight Perkins *et al, Rural Small–scale Industry in the People's Republic of China*, chapter V, pp. 117–18.)
14. The responsibility system is noted to have sharply reduced the demand for large-scale farm machinery and raised a market for custom services. See Thomas B. Wiens, 'Price Adjustment, the Responsibility System and Agricultural Productivity', *American Economic Review*, Papers and Proceedings (May 1983).
15. Shigeru Ishikawa, *Essays on Technology, Employment and Institutions in Economic Development*.
16. See Owen L. Dawson, *Communist China's Agriculture: Its Development and Future Potential* (New York: Praeger, 1970) p. 163.
17. Virgil A. Johnson and Halsey L. Beemer Jr (eds), *Wheat in the People's Republic of China* (Washington, D.C.: National Academy of Sciences, 1977).
18. Thomas B. Wiens, 'The Limits to Agricultural Intensification: The Suzhou Experience', in US Congress Joint Economic Committee, *China Under the Four Modernisations*, Part I (Washington, D.C., 1982) In this article, Wiens does not give detailed figures of actual mandays worked by crops and activity. Instead, he gives 'figures which are based on averages of reported work-day requirements' (in *gongs* equivalent to about 6 hours of work). The actual mandays figures quoted above are given in the text for only rapeseed and rice: no indication is given as to how they are

derived. It would appear that in estimating actual labour requirements in mandays (p. 470) Wiens is assuming an 8-hour day for rapeseed and a 10-hour day for single-crop rice.

19. Thomas G. Rawski, *Economic Growth and Employment in China* (Oxford University Press, 1979) p. 98.
20. See Frederic M. Surls and Francis C. Tuan, 'China's Agriculture in the Eighties', in *China Under the Four Modernisations*.
21. The estimation of labour costs in China is very different from that in a capitalist economy. The official estimates seem to be based on workpoints and remuneration in yuan per workpoint. The data shows that the labour cost per manday in cotton at 0.78 yuan is very close to the national average of 0.8 yuan for 1979.
22. Frederic M. Surls and Francis C. Tuan, 'China's Agriculture in the Eighties'.
23. 'Evolution and Development of the Paddy Crop Situation in the Lake District', *China Agricultural Bulletin* (*Chung-kuo nung-pao*) (1964) no. 8, p. 27.
24. Shigeru Ishikawa 'China's Food and Agriculture – A Turning Point', *Food Policy*, May 1977.
25. Benedict Stavis and Mitch Meisner, Introduction to 'China's Cropping System Debate', Special Issue of the *Chinese Economic Studies*, vol. xv (Winter 1981–2) no. 2.
26. Benedict Stavis and Mitch Meisner, *ibid.*; Thomas B. Wiens, 'The Limits to Agricultural Intensification: The Suzhou Experience'.
27. Benedict Stavis and Mitch Meisner, Introduction to 'China's Cropping System Debate'. Labour cost is based on an assumed wage rate of 0.8 yuan per day.
28. In many cases, peasant incomes are supplemented by such subsidiary activities as commune and brigade enterprises.

6 RURAL INDUSTRIALISATION

1. Peter Schran, 'Handicrafts in Communist China', *China Quarterly*, January–March 1964.
2. I am grateful to Shigeru Ishikawa for clarifying this point in a private communication.
3. These enterprises were frequently built in the small local towns. In 1980, the industrial output of the commune and brigade industries located in the small local towns was 57.9 per cent of their total output or 30.57 billion yuan. It may be noted that the small local towns are invariably classified as *urban*. Moreover, it is possible that even the county-run factories are also built in some of these towns.
4. This point is also due to Shigeru Ishikawa.
5. The county-run industry might sometimes be included and sometimes excluded from rural industry.
6. See Xu Kuan, Shen Xiaoli and Sun Fangming, 'On the Enterprises run by the People's Communes and Production Brigades', *Economic Theory and Business Management* (a bi-monthly) (The People's University of China,

1982) no. 1, pp. 41, 54–8 (in Chinese). I am grateful to my friend and former colleague, Chiang Hsieh, for translating this article into English.
7. A commune generally consists of 15 000 persons grouped into about 5000 households and 12–13 brigades. The size of a commune in terms of number of persons is known to vary between 5000 and 50 000. A production brigade consists of about 1600 persons grouped into 7–12 teams (some brigades are much smaller consisting of no more than 3–4 teams) and about 400 households. A production team consists of about 200 persons grouped into 20–50 households.
8. Xue Muqiao, 'On the Question of Commune and Brigade Enterprises', in *Several Problems Confronting Our National Economy* (*Danggian Woguo Jingji Yuegan Wenti*) (Beijing: People's Press, 1980) p. 113.
9. *Beijing Home Service*, 14 February 1981, in *BBC SWB* W 1123/A/4.
10. *NCNA*, 27 December 1981.
11. Xu Kuan, Shen Xiaoli and Sun Fangming, 'On the Enterprises run by the People's Communes and Production Brigades'.
12. *Changsha, Hunan Provincial Service*, 27 March 1981.
13. This has been an issue for a long time in China. The now purged 'Gang of Four' had argued that the bureaucrats always used this argument as an excuse to thwart autonomous local industrial activity because they wanted to keep control at the centre.
14. Shigeru Ishikawa, 'China's Food and Agriculture: Performance and Prospects', in Erwin M. Reisch (ed.), *Agriculture Sínica* (Berlin: Duncker & Humboldt, 1982).
15. This point is due to Shigeru Ishikawa.
16. The promotion of small town industrialisation is also desirable since the labour force can commute from the rural households where the workers were born and raised.
17. For example, currently Hunan uses prefabricated and cement blocks for building, for which specialised training is provided at the provincial level.
18. See Carl Riskin, 'Small Industry and the Chinese Model of Development', *China Quarterly* (1971) no. 46.
19. Li Yu and Chen Shengchang, 'Scale of Industrial Enterprises', *Social Sciences in China*, vol. II (June 1981) no. 2, p. 56.
20. There are 154 small paper mills in Hunan of which 59 are operated by communes and brigades. In 1980, there were only six medium and large paper mills. The Hunan Paper Company has been charged with the reorganisation of the existing small-scale mills which compete with the larger mills for raw materials.
21. 'Regulations on some Questions Concerning the Development of Enterprises run by Rural People's Communes and Production Brigades' (promulgated by the State Council 3 July 1979), in *Almanac of China's Economy in 1981*, compiled by the Economic Research Centre (Hong Kong: Modern Cultural Co. Ltd, 1982).
22. *FBIS*, August 1978.
23. *Changsha, Hunan Provincial Service*, November 1978, and *FBIS*, 17 November 1978.
24. 'Regulations on some Questions Concerning the Development of Enterprises run by Rural People's Communes and Production Brigades',

section VII on 'The Farming Out of Production by Urban Industries', p. 154.

25. This information is based on our interviews with the government officials in the Hunan Provincial Administration during November 1981.
26. The shareholders thus seem to get 15 per cent interest and a share of profits. It is, however, not clear whether profits are defined as surplus after 15 per cent interest is paid to shareholders.
27. The state budget mainly finances county-level enterprises.
28. 'Regulations on some Questions Concerning the Development of Enterprises run by Rural People's Communes and Production Brigades'.
29. State Council, 'Regulations Concerning some Problems in Developing Commune and Brigade Enterprises (Trial Draft)', *ZGJJNJ*, 1981.
30. 'Regulations on some Questions Concerning the Development of Enterprises run by Rural People's Communes and Production Brigades', p. 158.
31. *Ibid*., p. 156.
32. It is estimated that by mid-1960 'there were about 60 000 industrial units run by the counties (*xian*) for an average of about thirty such enterprises per county, and some 200 000 units run by the rural communes, not including the even smaller shops of the brigades'. See Carl Riskin, 'Intermediate Technology in China's Rural Industries', in Austin Robinson (ed.), *Appropriate Technology for Third World Development* (London: Macmillan, 1979). p. 54.
33. See Jon Sigurdson, 'The Changing Pattern of Inter-sectoral Relationships in the Rural Machinery Industry in China', in S. Watanabe (ed.), *Technology, Marketing and Industrialisation – Linkages Between Large and Small Enterprises* (New Delhi: Macmillan, 1983).
34. See Radha Sinha, 'Rural Industrialisation in China', in E. Chuta and S. Sethuraman (eds), *Rural Small-Scale Industries and Employment in Africa and Asia – A Review of Programmes and Policies* (Geneva: ILO, 1983).
35. Jon Sigurdson, 'Les Options Technologiques de la Chine d'Aujourd'hui', *Impact, Science et Société*, vol. XXIII (1973) no. 4, pp. 380–1.
36. 'Regulations on some Questions Concerning the Development of Enterprises run by Rural People's Communes and Production Brigades', p. 152.

7 CONCLUDING REMARKS

1. Li Yu and Chen Shengchang, 'Scale of Industrial Enterprises', *Social Sciences in China*, vol. II (June 1981) no. 2.
2. I am grateful to Joseph Stepanek for drawing my attention to this point in a private communication.
3. For fear of creating large-scale technological unemployment, the Indian government decided to subsidise 'cottage' industry even at the expense of large-scale industry. A 'common production programme' for cottage and large industry was introduced. This programme laid down protective and fiscal measures to ensure coexistence between the mechanised techniques

used in large industry and the labour-intensive techniques employed by the cottage industry, in sectors where competition existed.

4. For a treatment of some of these issues, see David C. Cole, 'Whether and How to Promote China's Exports', paper presented to the International Seminar on the Modernisation of Industry Related to Agriculture, Changsha, Hunan, 4–14 November 1981.
5. See Arthur G. Ashbrook, Jr, 'China: Economic Modernisation and Long-term Performance', in US Congress Joint Economic Committee, *China Under the Four Modernisations*, Part I (Washington, D.C., 1982) p. 109.

Bibliography

CHINESE SOURCES

Almanac of China's Economy in 1981 With Economic Statistics for 1949–1980, compiled by the Economic Research Centre, State Council of People's Republic of China and the State Statistical Bureau, ed. Xue Muqiao (Hong Kong: Modern Cultural Co. Ltd, 1982).

'Beijing's Pilot Project in Commerce – Events and Trends', *Beijing Review*, 31 January 1983.

Chen Weiqin, 'The Direction of Introducing Technology should be Changed', *JJGL* (15 April 1981) no. 4.

China Agricultural Yearbook (*Zhongguo Nongye Nianjian*), 1980, State Agricultural Commission.

China, People's Republic of: *Science and Technology for Turning China into a Prosperous and Strong Modern Socialist Country*, Report presented to the UN Conference on Science and Technology for Development (Beijing, 1979).

'China's Road to Agricultural Mechanisation', *Economic Reporter*, Hong Kong (1980) no. 12.

'China's Economy and Development Principles – A Report by Premier Zhao Ziyang', *Chinese Documents* (Beijing: Foreign Language Press, 1982).

'China's High Ranking Economic Official on the Chinese Investment Promotion Meeting', *Economic Reporter* (April 1982) no. 4.

'China's New Major Drive to Seek Foreign Investment', *Economic Reporter* (April 1982) no. 4.

Chu Xiangyin and Wang Shouchun, 'The Problem of Developing our Export of Labour-intensive Products', *International Trade Journal* (*Guoji Maoyi Wenti*) (Winter 1981) no. 4.

'Diversify the Rural Economy (Hunan Province)', *Beijing Review*, 7 September 1979.

'Economic Development in Hunan Province', in *Almanac of China's Economy in 1981* (Hong Kong: Modern Cultural Co. Ltd, 1982).

'Evolution and Development of the Paddy Crop Situation in the Lake District', *China Agricultural Bulletin* (*Chung-kuo nung-pao*) (1964) no. 8.

Feng Lanrui and Zhao Luküan, 'Urban Unemployment in China', *Social Sciences in China*, vol. III (March 1982) no. 1.

Gi Hong, 'Economic and Technical Cooperation in Agricultural Machinery between China and Foreign Countries', *Economic Reporter* (December 1980) no. 12.

Gu Nianliang, 'China's Current Effort to Import Technology and its

Prospects', *Chinese Economic Studies*, vol. XIV (Fall 1980) no. 1.
Gu Qianan, 'Study on the Use of Small Farm Machinery in Rice Production in Taihu Area, Jiangsu, China', paper prepared for the Seminar on Mechanisation of Small-scale Farming, Hangzhou, China, 22–6 June 1982.
He Jianzhang, 'Newly Emerging Economic Forms', *Beijing Review*, 25 May 1981.
Hsiang Jung and Chin Chi-chu, 'Change of the System of Ownership — Socialist Commerce (I) (Hunan Province)' *Peking Review* (9 July 1976) no. 28.
——, 'Develop the Economy and Ensure Supplies – Socialist Commerce (II) (Hunan Province)' *Peking Review* (23 July 1976) no. 30.
——, 'Not for Profits: Socialist Commerce (III) (Hunan Province)' *Peking Review* (30 July 1976) no. 31.
——, 'A Vast Rural Market: Socialist Commerce (IV)' (Hunan Province) *Peking Review* (9 August 1976) nos. 32–3.
Hu Mengzhou, 'Solution to Employment Problems', *Beijing Review*, 27 September 1982.
Hua Guozhu and Yao Jianfu, 'Some Aspects and Experience in China's Agricultural Mechanisation', paper prepared for the Seminar on Mechanisation of Small-scale Farming, Hangzhou, China, 22–6 June 1982.
'Hunan Issues Circular on Agricultural Economic Policies', *Changsha, Hunan Provincial Service*, 29 February 1980.
Hunan Province, Translation from the Chinese Press and Radio Broadcasts (Washington, D.C.: Library of the National Council for US–China Trade, 10 August 1981).
Hung Yuanpeng and Weng Qiquan, 'On Urban Collective Industries', *JJYJ* (1980) no. 1.
'Industry, Transportation Conference held in Hunan', *Changsha, Hunan Provincial Service*, 4 April 1981.
'It is Essential to Attach Importance to Farm Machinery Management', *People's Daily*, 18 March 1980, translated in *SWB* (FE/W 1079/A/10), 23 April 1980.
'It is Imperative to Relentlessly Grasp Scientific and Technological Work', *Hunan Ribao*, 14 July 1980.
'Joint Operations in Light Industry', *Peking Service*, 23 May 1980. *SWB*, FE 1085/A/10, 4 June 1980.
Li Chengrui and Zhang Zhuoyuan, 'Comprehensive Planning of Employment and Making Full Use of Manpower Resources', JJYJ (1980) no. 8.
Li Yu and Chen Shengchang, 'Scale of Industrial Enterprises', *Social Sciences in China*, vol. II (June 1981) no. 2.
Liang Chuanyuan, 'Problems Concerning the Recovery and Development of Individual Industrialists and Businessmen', *JJGL* (1980) no. 7.
Liang Liguang, 'Develop Regional Superiority, Greatly Develop Light Industry – A Survey of the Light Industries in Hubei and Hunan', *Congren Ribao* (*Workers' Daily*), Beijing, 17 July 1980.
Liang Wensen, 'Balanced Development of Industry and Agriculture', in Xu Dixin *et al.* (eds), *China's Search for Economic Growth — The Chinese Economy since 1949*, China Studies Series (Beijing: New World Press, 1982).

Liao Jianxiang, 'Size of Industrial Enterprise Operation and Choice of Technology', in Xu Dixin *et al.* (eds), *China's Search for Economic Growth – The Chinese Economy since 1949*, China Studies Series (Beijing: New World Press, 1982).

Liao Jili, 'A Discussion of Organisational Change From the Viewpoint of Horizontal and Vertical Relations', *RMRB*, 26 August 1980.

Lin Zili, 'Initial Reform in China's Economic Structure', *Social Sciences in China*, vol. I (September 1980) no. 3.

Liu Wenduo, 'Interplanting and Intercropping System in North China and Adaptability of 5-Horsepower Unit for it', paper prepared for the Seminar on Mechanisation of Small-scale Farming, Hangzhou, 22–6 June 1982.

Liu Zhongyi and Lian Yaochuan, 'Rural Economic Structure and its Relation to Agricultural Mechanisation in China', paper prepared for the Seminar on Mechanisation of Small-scale Farming, Hangzhou, 22–6 June 1982.

Lu Zhangshan, 'Develop Xian and Commune Enterprises to Accelerate the Realisation of Agricultural Mechanisation', *Economic Research* (1978) no. 5.

Lui Wenwei, 'Commune and Brigade Enterprises in Hunan', paper presented at the International Seminar on the Modernisation of Industry Related to Agriculture, Changsha, Hunan, 4–14 November 1981.

Mao Tse Tung, 'On the Ten Major Relationships', *Peking Review*, 1 January 1977.

'More Efficient Use of Fiscal Farming Aid Funds', *Economic Research* (20 February 1980) no. 2, translated in SWB, FE/W 1082/A/7, 14 May 1980.

'New Experimental Farms (Hunan and other Provinces)', *Beijing Review*, 26 January 1979.

'Open Policy Hand in Hand With Protectionism – A Paradox?', by a Staff Correspondent, *Economic Reporter* (May 1982) no. 5.

Peng Xianchu, 'Rural Responsibility System: Spot Report (on Hunan), Part I', *China Reconstructs*, vol. XXXI (August 1982) no. 8.

Qi Zong, 'The Commune and Brigade Industries in Rural China', *Annual Economic Report IV*, 1981.

'Reform of the Employment System', *Beijing Review*, 4 April 1983.

'Regulations on some Questions Concerning the Development of Enterprises run by Rural People's Communes and Production Brigades' (promulgated by the State Council, 3 July 1979), in *Almanac of China's Economy in 1981*, compiled by the Economic Research Centre (Hong Kong: Modern Cultural Co. Ltd, 1982).

Ren Jianxin, 'Legal Aspects of China's Technology Import and Utilisation of Foreign Investment', *Economic Reporter* (September 1980) no. 9.

State Council, 'Regulations Concerning some Problems in Developing Commune and Brigade Enterprises (Trial Draft)', *ZGJJNJ* (Annual Economic Report of China), 1981.

Statistical Yearbook of China for 1981 (People's Republic of China: SSB, 1982).

'Strengthening Centralisation and Unity is the Key to Making a Success of Economic Readjustment', *Changsha, Hunan Provincial Service*, 28 February 1981.

Sung Tao, 'On China's New Economic Policies', *China Enterprise*, Hong Kong, December 1981.

Teng Weizao, 'Socialist Modernisation and the Pattern of Foreign Trade', in Xu Dixin *et al.* (eds), *China's Search for Economic Growth – The Chinese Economy since 1949*, China Studies Series (Beijing: New World Press, 1982).

Wang Bingqian, 'Report on Financial Work', *Beijing Review*, 29 September 1980.

Wang Enkui, 'A Discussion of Technological Economics', *HQ* (1980) no. 7.

Wang Leo-bao and Lin Qi-hui, 'The Development and Prospect of Agricultural Machinery Industry and Agricultural Mechanisation in China', paper prepared for the Seminar on Mechanisation of Small-scale Farming, Hangzhou, 22–6 June, 1982.

Wang Shoudao, 'The Responsibility Systems have brought New Vitality to the Rural Areas of Hunan – Report of an Investigation of Several Prefectures and Counties in Hunan', *RMRB* Beijing, 5 June 1981.

Wang Wenke, 'Trade Marks in China', *Economic Reporter* (April 1982) no. 4.

Wang Yaoling 'Organisation of Foreign Trade in China', paper presented at the International Seminar on the Modernisation of Industry Related to Agriculture, Changsha, Hunan, 4–14 November, 1981.

'We Must Get a Good Grasp of Grain Production', *Changsha, Hunan Provincial Service*, 24 January 1981.

'We Must Pay Particular Attention to Consumption Growth', *Trends in Economic Studies* (*Jingjixue Dongtai*) (22 June 1982) no. 6.

Wong Yongxi *et al.*, 'Views on Strategic Problems in China's Agricultural Development', *JJYJ*, November 1981.

Wu Jiuxin, 'Beijing is Solving its Youth Unemployment Problem', *China Reconstructs*, vol. XXXII, June 1983, no. 6.

Xia Deji, 'Agricultural Mechanisation in Hunan', paper presented at the International Seminar on the Modernisation of Industry Related to Agriculture, Changsha, Hunan, 4–14 November 1981.

Xia Yibai, 'Acquisition of Technology for the Modernisation of Machine-building in Hunan', paper presented at the International Seminar on the Modernisation of Industry Related to Agriculture, Changsha, Hunan, 4–14 November 1981.

Xia Zhen, 'New Strategy for Economic Development', *Beijing Review*, 10 August 1981.

Xia Zhen-kun, Wang Wen-Lung, Zhou Yong-Quan and Chen Zhen-Gao, 'The Adaptability and Economy of Boat Tractor', paper prepared for the Seminar on Mechanisation of Small-scale Farming, Hangzhou, 22–6 June 1982.

Xu Kuan, Shen Xiaoli and Sun Fangming, 'On the Enterprises run by the People's Communes and Production Brigades', *Economic Theory and Business Management* (The People's University of China, Beijing, 1982) no. 1.

Xue Muqiao, *China's Socialist Economy* (Beijing: Foreign Language Press, 1981).

——, 'Addendum to "China's Socialist Economy"', *Beijing Review*,

December 1981.
——, 'On the Question of Commune and Brigade Enterprises', in *Several Problems Confronting Our National Economy* (*Danggian Woguo Jingji Yuegan Wenti*) (Beijing: People's Press, 1980).
——, 'China's Economic Readjustment: text of the 8 May Speech at the First Sino–Japanese Meeting for the Exchange of Economic Knowledge', *Shanghai 'Shijie Jingji Daobao'* (1 June 1981) no. 35, translated in *SWB*, FE/W 1140/C/1, 1 July 1981.
——, 'Tentative Study on the Reform of the Economic System', *Chinese Economic Studies*, vol. xiv (Winter–Spring 1980–1) nos. 2–3 (translated from *JJYJ*).
Yang Jianbai and Li Xuezeng, 'The Relations between Agriculture, Light Industry and Heavy Industry in China', *Social Sciences in China*, vol. i, (June 1980) no. 2.
Yao Yilin, 'Report on the Adjustment of the 1981 National Economic Plan and State Revenue and Expenditure (Excerpts)', *Beijing Review*, 16 March 1981.
You Yuwen, 'Jobless Youth Start Small Businesses', *China Reconstructs*, vol. xxx (October 1981) no. 10.
Zeng Dechao, 'Rural Development and Appropriate Technology in China', paper presented at the International Seminar on the Modernisation of Industry Related to Agriculture, Changsha, Hunan, 4–14 November 1981.
Zeng Qixian, 'Employment Creation and Economic Development', in Xu Dixin *et al.* (eds), *China's Search for Economic Growth – The Chinese Economy since 1949*, China Studies Series (Beijing: New World Press, 1982).
Zhai Yao, 'Problems in the Mechanisation of China's Agriculture', *Economic Reporter* (July 1980) no. 7.
Zhan Wu, 'Take the Road of Agricultural Modernisation the Chinese Way', *JJGL* (1979) no. 9; English translation in *Chinese Economic Studies*, vol. xiv (Summer 1981) no. 4.
Zhan Wu, He Naiwei and Zhang Baomin, 'The Farm Responsibility System has Unfolded the Merits of the Collective Agriculture of Socialism', *HQ* (1981) no. 17.
Zhang Shugyuang, 'Reform the Economic Structure and Increase the Macroeconomic Results – An Analysis of China's Economic Structure and the Changes in Economic Results', *Social Sciences in China*, vol. iii (March 1982) no. 1.
Zhang Yulin, Yang Chengxun and Guo Xiping, 'Contracting Output Quotas to Households Under Unified Management: The Responsibility System in the Xiaotan People's Commune (Henan province)', *Social Sciences in China*, vol. iv (June 1983) no. 2.
Zhou Jizhong and Chen Cangye, 'We Should Establish Development Research Centres', *Guangming Ribao*, 1 August 1980.
Zhu Lunkun *et al.*, 'Interaction between Farm Mechanisation and Commune-run Industry in Shanghai Rural Areas', paper prepared for the Seminar on Mechanisation of Small-scale Farming, Hangzhou, 22–6 June 1982.

OTHER SOURCES

Ahmad Aqueil, 'Science and Technology in Development: Policy Options for India and China', *Economic and Political Weekly*, Bombay, vol. XIII (23–30 December 1978) nos. 51 and 52.

——, 'Flows of Science and Technology Information: The Cases of India and China', paper presented to the Conference on Communication Support for Rural Development, Administrative Staff College of India, Hyderabad, 27–30 December 1978.

Aird, John S., *Population Estimates for the Provinces of the People's Republic of China, 1953–1974*, (Washington, D.C.: US Department of Commerce, February 1974).

Ashbrook, Arthur G. Jr 'Economic Modernisation and Long-term Performance, in US Congress Joint Economic Committee, *China Under Four Modernisations*, Part I.

Baark, Erik, *Dissemination of Technology Information in China: An Investigation in Publishing in Electronics and Metallurgy*, Discussion Paper no. 127 (Research Policy Institute, Lund University, 1979).

——, *Techno-economics and Politics of Modernisation in China: Basic Concepts of Technology Policy under the Readjustment of the Chinese Economy*, Discussion Paper no. 135 (Research Policy Institute, Lund University, November 1980).

——, 'China's Technological Economics', *Asian Survey*, September 1981.

Balassa, Bela, 'Economic Reform in China', *Banca Nazionale del Lavoro* (Quarterly Review) Rome, September 1982.

Barker, R. *et al* (eds), *The Chinese Agricultural Economy*, Westview Special Studies on China and East Asia (Boulder, Colorado: Westview Press, 1982).

Barnett, A. D., *China in a Global Perspective* (Washington, D. C.: Brookings Institution, 1981).

Baum, R. (ed.), *China's Four Modernisations: The New Technological Revolution* (Boulder, Colorado: Westview Press, 1980).

Berner, B., 'The Organisation and Planning of Scientific Research in China Today', Discussion Paper no. 134 (Research Policy Institute, Lund University 1979).

Bhalla, A. S., 'Technological Choice in Construction in Two Asian Countries: China and India', *World Development*, March 1974.

——, *Rural Industrialisation and New Economic Policies in Hunan (China)*, ILO/WEP Research Working Paper Series, WEP 2–37, no. 9, Geneva (May 1982).

——, 'Internal Technology Transfers in China', paper presented at the International Conference on Science and Technology Policy and Research Management, organised by the Chinese State Science and Technology Commission in conjunction with the UN Financing System on Science and Technology for Development (UNFSSTD), Beijing (4–8 October 1983).

Bhalla, A.S. and Edmonds, G.A., 'Construction Growth and Employment in Developing Countries', *Habitat International*, Oxford, vol. 7 (1983) nos 5–6.

Blecher, Marc, 'Urban and Rural in a Developing Chinese County — The

Case of Shulu', paper prepared for the Conference on China in Transition, Queen Elizabeth House, Oxford, 7–10 September 1982.

Brugger, Bill (ed.), *China Since the Gang of Four* (London: Croom Helm, 1980).

——, 'Rural Policy', in *China Since the Gang of Four* (London: Croom Helm, 1980).

Burki, Shahid Javed, *A Study of Chinese Communes* (Cambridge, Mass.: Harvard University Press, 1969).

Chakrabarti, Sreemati, 'China's Hunan Province', *China Report*, New Delhi, vol. XVIII (July–August 1982) no. 4.

Chao, Kang, *The Construction Industry in Communist China* (Edinburgh University Press, 1968).

Chudnovsky, Daniel and Masafumi Nagao, *Capital Goods Production in the Third World — An Economic Study of Technology Acquisition* (London: Frances Pinter, 1983).

Chu-yuan Cheng, *The Machine-Building Industry in Communist China* (Chicago: Aldine, 1971).

——, *China's Economic Development: Growth and Structural Change* (Boulder, Colorado: Westview Press, 1982).

Clarke, Christopher, M., *China's Provinces: An Organisational and Statistical Guide* (Washington, D.C.: National Council for US–China Trade, 1982).

Cole, David C., 'Whether and How to Promote China's Exports', paper presented at the International Seminar on the Modernisation of Industry Related to Agriculture, Changsha, Hunan, 4–14 November 1981.

Conroy, Richard, 'Supply and Demand for Technological Innovations in China's Present Phase of Industrial Modernisation', paper prepared for the Conference on China in Transition, Queen Elizabeth House, Oxford University, 7–10 September 1982.

Current Literature on Science of Sciences, National Institute for Science, Technology and Development Studies (New Delhi, March 1982).

Datta, Rakhal, 'Technology Choice in Collectivised Agriculture: Farm Mechanisation Policy of the People's Republic of China', Special Articles, *China Report*, New Delhi, vol. XVI (September–October 1980) no. 5.

Dawson, Owen L., *Communist China's Agriculture: Its Development and Future Potential* (New York: Praeger 1970).

Dean, G. C., *Science and Technology Policy for Development: Technology Policy and Industrialisation in the People's Republic of China*, IDRC, STPI, no. 4 (Ottawa, 1979).

Dernberger, R. F., 'Economic Development and Modernisation in Contemporary China: The Attempt to Limit Dependence on the Transfer of Modern Industrial Technology from Abroad and to Control its Corruption of the Maoist Socialist Revolution', in Frederic J. Fleron, Jr (ed.), *Technology and Communist Culture* (New York: Praeger, 1977).

——, *China's Development Experience in Comparative Perspective* (Cambridge, Mass.: Harvard University Press, 1980).

——, 'The Chinese Search for the Path of Self-Sustained Growth in the 1980's: An Assessment', in US Congress Joint Economic Committee, *China Under the Four Modernisations*, Part I (Washington, D.C., August 1982).

Dickinson, H., 'The Role of Technical and Industrial Education in the

Development of China', *Occasional Papers on Appropriate Technology* (Edinburgh, March 1977).

Ding, Chen, 'The Economic Development of China', *Scientific American*, vol. 243 (September 1980) no. 3.

Dobb, M., *An Essay on Economic Growth and Planning* (Modern Reader, 1960).

Domes, Jürgen, 'New Policies in the Communes: Notes on Rural Societal Structures in China, 1976–1981', *Journal of Asian Studies*, vol. XLI (February 1982) no. 2.

Eckstein, Alexander, *China's Economic Revolution* (Cambridge University Press, 1977).

ESCAP, 'Report of the Workshop on Small and Medium-scale Industries at Selected Sites in the People's Republic of China (21 October–4 November 1978)', *Small Industry Bulletin for Asia and the Pacific*, no. 16 (New York: United Nations, 1979).

Etienne, G., *La Chine Fait Ses Comptes*, IEDES Section on 'Quelques données sur le Hunan' (Paris: Collection Tiers Monde, Presses Universitaires de France, 1980).

——, 'Industries et énergie en Chine: des options délicates', *Revue Tiers–Monde*, April–June 1981.

Field, Robert Michael, 'Real Capital Formation in the People's Republic of China, 1952–1973', April 1976 (unpublished manuscript).

Field, R., Lardy, N. and Emerson, J., *Provincial Industrial Output in the People's Republic of China, 1957–75* (Washington, D.C.: US Department of Commerce, 1976).

'Foreigners are Dangerous – Official', *Far Eastern Economic Review*, 1–7 October 1982.

Fujimoto, Akira, 'The Economic Responsibility System in China's Industrial Sector', *China Newsletter* (Tokyo, JETRO) (November–December 1982) no. 41.

Ghose, Ajit Kumar, *The New Development Strategy and Rural Reforms in Post-Mao China*, WEP Working Paper WEP 10–6/WP. 62 (ILO, Geneva, November 1983).

Godement, F., 'Financer le Développement ou l'Accumulation?', *Revue Tiers–Monde*, April–June 1981.

Gray, J., *Rural Industrialisation in China, 1977–1979* (Brighton: Institute of Development Studies, University of Sussex, 1979).

Griffin, Keith and Griffin, Kimberley, 'Institutional Change and Income Distribution in the Chinese Countryside', *Oxford Bulletin of Economics and Statistics*, vol. 45, no. 3 (August 1983).

Griffin, K. and A. Saith, *Growth and Equality in Rural China* (Singapore: ILO, Asian Employment Programme (ARTEP), 1981).

Gurley, J., *China's Economy and the Maoist Strategy* (New York: Monthly Review Press, 1976).

Hama, Katsuhiko, 'China's Agricultural Production Responsibility System', *China Newsletter* (Tokyo) (September–October 1982) no. 40.

Hawkins, J.N., 'Rural Education and Technique Transformation in the People's Republic of China', *Technological Forecasting and Social Change* (Elsevier, North-Holland) (1978) no. 11.

Heyden, Andrew, 'The Modern Ceramics Trade', *The China Business Review* (Washington, D.C.) September–October 1982.

Hollister, William W., 'Trends in Capital Formation in Communist China', in *An Economic Profile of Mainland China*, vol. I, Studies prepared for the US Congress Joint Economic Committee (Washington D.C., February 1967).

Hsu, Robert C., 'Agricultural Mechanisation in China: Policies, Problems and Prospects', *Asian Survey*, vol. XIX (May 1979) no. 5.

——, *Food For One Billion – China's Agriculture since 1949* (Boulder, Colorado: Westview Press, 1982).

Ishikawa, Shigeru, 'A Note on the Choice of Technology in China', *Journal of Development Studies*, vol. 9 (October 1972) no. 1.

——, 'The Chinese Method of Technological Development: The Case of Agricultural Machinery', *The Developing Economies*, vol. XIII (December 1975) no. 4.

——, 'China's Food and Agriculture – A Turning Point', *Food Policy*, May 1977.

——, *Prospects of the Chinese Economy in the 1980s* (Tokyo: Institute of Economic Research, Hitotsubashi University, 1979).

——, *A Note on China's Economic Performance, 1949–1979*, paper presented to the World Bank, 2 September 1980, mimeo.

——, *Essays on Technology, Employment and Institutions in Economic Development*, Comparative Asian Experience, Economic Research Series no. 19 (Tokyo: Institute of Economic Research, Hitotsubashi University, Kinokuniya Company Ltd, 1981).

——, 'China's Food and Agriculture: Performance and Prospects', in Erwin M. Reisch (ed.), *Agriculture Sinica* (Berlin: Duncker & Humbolt, 1982).

——, 'China's Economic Growth in the PRC Period — An Assessment', *China Quarterly* (June 1983), no. 94.

Ishikawa, Shigeru, Saburo Yamada and S. Hirashima, *Labour Absorption and Growth in Agriculture – China and Japan* (Singapore: ILO, Asian Employment Programme (ARTEP), 1982).

Johnson, E. and Johnson, G., *Walking on Two Legs, Rural Development in South China* (Ottawa: IDRC, 1976).

Johnson, Graham E., 'The Production Responsibility System in Chinese Agriculture: Some Examples from Guang dong', *Pacific Affairs*, vol. 55 (Fall 1982) no. 3.

Johnson, Virgil A. and Halsey Beemer Jr (eds), *Wheat in the People's Republic of China* (Washington, D.C.: National Academy of Sciences, 1977).

Jowett, J.A., 'China: The Provincial Distribution of Population', *China Quarterly*, March 1980.

Keesing, Donald B., 'Economic Lessons from China', *Journal of Development Economics*, 2, (Amsterdam, North-Holland, 1975).

Khan, Amir U., 'Agricultural Mechanisation and Machinery Production in China', *Agricultural Mechanisation in Asia*, Spring 1976.

Khan, Azizur Rahman, 'The Responsibility System and the Change in Rural Institutions in China' in Khan, Azizur Rahman and Lee, Eddy, *Agrarian Policies and Institutions in China after Mao*, ILO/Asian Employment Programme (ARTEP) (Maruzen, Singapore, 1983).

—— and Lee, Eddy, *Agrarian Policies and Institutions in China after Mao*, ILO–Asian Employment Programme (ARTEP) (Singapore: Maruzen, 1983).

Kojima, R., 'China's New Agricultural Policy', *Developing Economies*, Tokyo, vol. xx (December 1982) no. 4.

Kueh, Y. Y: *Local Level Planning in China*, ILO/WEP Research Working Paper Series No. WEP 2–32/WP. 44 (Geneva, November 1982).

Lalkaka, Dinyar, *Urban Housing in China* (Beijing, June 1983, mimeo.).

Lalkaka, R., *Small and Medium-scale Industries in China*, Report of a UNIDO Study Tour (6–30 October 1977), UNIDO/IOD, 172, 29 March 1978.

Lele, Uma., 'Rural Marketing in China: A Comparative Perspective', *World Development*, vol. 6 (May 1978) no. 5.

Lippit, Victor and Mark Selden (eds), *The Transition to Socialism in China* (White Plains, N.Y.: M. E. Sharpe, 1982).

Lardy, Nicholas R, *Economic Growth and Distribution in China* (Cambridge University Press, 1978).

——, *Agriculture in China's Modern Economic Development* (Cambridge University Press, 1983).

Macrae, John T., 'A Clarification of Chinese Development Strategy since 1949', *Developing Economies*, vol. xvii (September 1979) no. 3.

Maier, John H., 'Information Technology in China', *Asian Survey*, vol. 20 (August 1980) no. 8.

Maruyama, Nobuo, 'The Mechanism of China's Industrial Development – Background to the Shift in Development Strategy', *Developing Economies*, vol. xx (December 1982).

Modernisation of Industry Related to Agriculture in Hunan (China), Papers and Proceedings of the Changsha International Seminar (Changsha, 4–14 November 1981) (VITA, US), 1983.

Morawetz, David, 'Walking on Two Legs? Reflections on a China Visit', *World Development*, vol. 7 (August–September 1979) nos. 8–9.

Morehouse, Ward, 'Notes on Hua-Tung Commune (Commune as a Technological System)', *China Quarterly* (September 1976) no. 67.

Myers, R., *The Chinese Economy: Past and Present* (Belmont, California: Wadsworth, 1980).

Nai-Ruenu Cheu, 'Economic Modernisation in Post-Mao China: Policies, Problems and Prospects', in *Chinese Economy Post-Mao – A Compendium of Papers submitted to the Joint Economic Committee of US Congress*, vol. i (Washington, D.C., November 1978).

Nolan, Peter, 'Decollectivisation of Agriculture in China 1979–1982: A Long-term Perspective', *Economic and Political Weekly*, 6 and 13 August 1983.

OECD, *Science and Technology in the People's Republic of China* (Paris, 1977).

Onoye, E., 'Readjustment and Reform in the Chinese Economy: A Comparison of the Post-Mao and Post-Great Leap Forward Periods', *Developing Economies,* December 1982.

Orleans, Leo A., 'China's Urban Population: Concepts, Conglomerations, and Concerns', in US Congress Joint Economic Committee, *China Under the Four Modernisations,* Part I (Washington D.C., 1982).

Pairault, T., 'Travail rural et rémunération en Chine: quelques aspects de la politique economique rural', *Tiers-Monde*, Paris (October–December 1979).

Perkins, Dwight *et al.*, *Rural Small-Scale Industry in the People's Republic of China* (The American Rural Small-scale Industry Delegation, University of California Press, 1977).

Rawski, T. G., *Economic Growth and Employment in China* (Oxford University Press, 1979).

——, *China's Transition to Industrialism: Producer Goods and Economic Development in the Twentieth Century* (Ann Arbor, Michigan: University of Michigan Press, 1980).

——, 'Choice of Technology and Technological Innovation in China's Economic Development', in Robert F. Dernberger (ed.), *China's Development Experience in Comparative Perspective* (Cambridge, Mass.: Harvard University Press, 1980).

Reynolds, Bruce L., 'Reform in Chinese Industrial Management: An Empiricial Report', in US Congress Joint Economic Committee, *China Under the Four Modernisations*.

Rice Research and Production in China: An IRRI Team's View (Philippines: International Rice Research Institute, 1978).

Riskin, C., 'Local Industry and the Choice of Technique in the Planning of Industrial Development in Mainland China', in UNIDO, *Planning for Advanced Skills and Technologies*, Industrial Planning and Programming Series no. 3 (New York, 1969).

——, 'Small Industry and the Chinese Model of Development', *China Quarterly* (1971) no. 46.

——, 'China's Rural Industries: Self-reliant Systems or Independent Kingdoms?', *China Quarterly*, March 1978.

——, 'Political Conflict and Rural Industrialisation in China', *World Development*, vol. 6 (1978).

——, 'Intermediate Technology in China's Rural Industries', in Austin Robinson (ed.), *Appropriate Technology for Third World Development* – Proceedings of a Conference held by the International Economic Association (London: Macmillan, 1979).

——, 'Market, Maoism, and Economic Reform in China', in Victor Lippit and Mark Selden (eds), *The Transition to Socialism in China* (White Plains, N.Y.: M. E. Sharpe, 1982).

——, *The Terms of Trade Between Industry and Agriculture in China* (1983, mimeo.).

Schran, P., 'Handicrafts in Communist China', *China Quarterly*, January–March 1964.

Sen, Amartya, *Employment, Technology and Development* (Oxford: Clarendon Press, 1975).

Seybolt, Peter J., *The Rustication of the Urban Youth in China: A Social Experiment* (White Plains, N.Y.: M. E. Sharpe, 1977).

Shirk, Susan L., 'Recent Chinese Labour Policies and the Transformation of Industrial Organisation in China', *China Quarterly* (December 1981) no. 88.

Sigurdson, Jon, 'Rural Industry – A Traveller's View', *China Quarterly*, April–June 1972.

——, 'Les Options Technologiques de la Chine d'Aujourd–hui', *Impact, Science et Société*, vol. XXIII (1973) no. 4.

——, 'Rural Industry and the Internal Transfer of Technology', in Stuart Schram (ed.), *Authority, Participation and Cultural Change in China* (Cambridge University Press, 1973).

——, 'Rural Industrialisation in China', *World Development*, July–August 1975.

——, *Rural industrialisation in China* (Cambridge, Mass.: Harvard University Press, 1977).

——, 'Technology and Science: Some Issues in China's Modernisation', in *Chinese Economy Post-Mao – A Compendium of Papers submitted to the Joint Economic Committee of US Congress*, vol. I (Washington, D.C., November 1978).

——, *China's Road to Autonomy in Technology and Science: Some Preliminary Observations* (Research Policy Institute, Lund University, 1979).

——, *Technology and Science in the People's Republic of China – An Introduction* (Oxford: Pergamon Press, 1981).

——, 'The Changing Pattern of Inter–sectoral Relationships in the Rural Machinery Industry in China', in S. Watanabe (ed.), *Technology, Marketing and Industrialisation – Linkages Between Large and Small Enterprises* (New Delhi: Macmillan, 1983).

Sigurdson, J. and B. Billgren, 'An Estimate of Research and Development Expenditure in the People's Republic of China in 1973', *Industry and Technology Occasional Paper no. 16* (Paris: OECD Development Centre, July 1977).

Simon, Denis Fred, 'China's Capacity to Assimilate Foreign Technology – An Assessment', in *China under the Four Modernisations*, Part I (Washington, D.C., 1982).

Sinha, Radha, 'Rural Industrialisation in China', in E. Chuta and S. Sethuraman (eds), *Rural Small-scale Industries and Employment in Africa and Asia – A Review of Programmes and Policies* (Geneva: ILO, 1983).

Solinger, D., 'The Shadowy Second Stage of China's Ten–Year Plan: Building Up Regional Systems, 1976–1985', *Pacific Affairs*, Summer 1979.

Stavis, Benedict, 'Rural Institutions in China', in R. Barker *et al.* (eds), *The Chinese Agricultural Economy*, (Boulder, Colorado: Westview Press, 1982).

Stavis, Benedict and Mitch Meisner, 'China's Cropping System Debate', Introduction, *Chinese Economic Studies*, Special Issue vol. XV (Winter 1981–2) no. 2.

Stepanek, James B., 'Planning of Urban Small-scale Industry in China', paper presented at a Conference sponsored by the Sub-Committee on Research on the Chinese Economy, Social Science Research Council, on 'Regionalism and Economic Development in China: Historical and South Asian Comparative Perspectives', (Philadelphia, 20–21 January 1978).

Surls, F. M. and F. C. Tuan, 'China's Agriculture in the Eighties', in *China under the Four Modernisations*, Part I (Washington, D.C., 1982).

Suttmeier, R.P., *Science, Technology and China's Drive for Modernisation* (Stanford University, California: Hoover International Studies, 1980).

Tell, M., *A Note About China's Scientific and Technological Information System*, Report TRITA–LIB 1102, Stockholm Papers in Library and Information Science (Stockholm: Royal Institute of Technology Library, 1980).

United Nations, *Trade Statistics, 1970–1979* (New York, 1980).

US Congress Joint Economic Committee, *China Under the Four Modernisations*, Part I (Washington D.C., 1982).

Wang, George C. (ed.), *Economic Reform in the People's Republic of China*, Westview Special Studies (Boulder, Colorado: Westview Press, 1982).

Watson, Andrew, 'Industrial Development and the Four Modernisations', in Bill Brugger (ed.), *China Since the Gang of Four* (London: Croom Helm, 1980).

Weiss, U., 'China's Rural Marketing Structure', *World Development*, vol. 6 (May 1978) no. 5.

Weisskopf, T. E., 'The Relevance of the Chinese Experience for Third World Economic Development', *Theory and Society*, March 1980.

Wheelwright, E. L. and McFarlane, B., *The Chinese Road to Socialism* (New York: Monthly Review Press, 1970; Harmondworth: Penguin, 1973).

Wiens, Thomas B., 'The Limits to Agricultural Intensification: The Suzhou Experience', in *China Under the Four Modernisations*, Part I (Washington, D.C., 1982).

——, 'Price Adjustment, the Responsibility System and Agricultural Productivity', *American Economic Review, Papers and Proceedings* (May 1983).

Wong, Christine Pui Wah, 'Rural Industrialisation in the People's Republic of China: Lessons from the Cultural Revolution Decade', in *China under the Four Modernisations*, Part I (Washington, D.C., 1982).

Wong, John, *Labour Mobilisation in the Chinese Commune System: A Perspective from Guangdong*, Asian Employment Programme Working Papers, ILO (Bangkok, January 1982).

World Bank, *World Development Report for 1982* (Washington, D.C., Oxford University Press and World Bank, 1982).

——, *China: Socialist Economic Development*, 3 vols (Washington, D.C., August 1983).

Wu, I–Chuan, 'Nourrir le Peuple: Politique Alimentaire et Politique Démographique en Chine', *Revue Tiers–Monde*, vol. XXII (April–June 1981) no. 86.

Zhang, Yan, 'Some Problems Concerning Technological Policy of Housing in China', in Ove Granstrand and Jon Sigurdson (eds), *Technological and Industrial Policy in China and Europe*, Proceedings from the First Joint TIPCE Conference, Lund, 1981.

Name Index

The index refers also to the bibliography which includes contributions relevant to the book but not specifically cited in the text.

Ahmad, Aqueil, 183
Aird, John S., 138, 183
Ashbrook, Arthur G. Jr, 177, 183

Baark, Erik, 75, 169–70, 183
Balassa, Bela, 183
Barker, R., 168, 183
Barnett, A.D, 183
Baum, Richard, 165, 183
Beemer, Halsey L. Jr., 173, 186
Berner, B., 183
Bhalla, A.S., 165, 170, 172, 183
Billgren, Boel, 71, 170, 189
Blecher, Marc, 183–4
Brugger, B., 165, 167, 184, 190
Burki, Shahid Javed, 184

Chakrabarti, Sreemati, 184
Chao, Kang, 171, 184
Chen, Cangye, 170, 182
Chen, Shengchang, 123, 175–6, 179
Chen, Weiqin, 84, 92, 171–2, 178
Chen, Zhen-Gao, 181
Cheng, Chu-yuan, 184
Chin, Chi-chu, 26, 28, 165, 179
Chu, Xiangyin, 178
Chudnovsky, Daniel, 170, 184
Chuta, E., 176, 189
Clarke, Christopher M., 82, 171, 184
Cole, David C., 177, 184
Conroy, Richard, 184

Datta, Rakhal, 172, 184
Dawson, Owen L., 106, 173, 184
Dean, Genevieve C., 60, 169, 184
Deng, Xiaoping, i, 45,
Dernberger, Robert F., 166, 169, 171, 184, 188
Dickinson, H., 184–5
Ding, Chen, 185
Dobb, M., 58, 169, 185
Domes, Jürgen, 45, 167–8, 185

Eckstein, Alexander, 57, 167, 169, 185
Edmonds, G.A., 165, 183
Emerson, J., 185
Etienne, G., 185

Feng, Lanrui, 168, 178
Field, Robert M., 171, 185
Fleron, Frederic, J. Jr, 184
Fujimoto, Akira, 168, 185

Ghose, Ajit Kumar, 166, 185
Gi, Hong, 178
Godement, F., 185
Gray, J., 185
Granstrand, Ove, 190
Griffin, Keith, 167, 185
Griffin, Kimberley, 185
Gu, Nianliang, 178–9
Gu, Qianan, 179
Guo, Xiping, 182
Gurley, J., 185

Hama, K., 44, 46, 167–8, 185
Hawkins, John N., 170–1, 185
He, Jianzhang, 166, 179
He, Naiwei, 182
Heyden, Andrew, 165, 186
Hirashima, S., 186
Hollister, William, 169, 186
Hsiang, Jung, 26, 28, 165, 179
Hsieh, Chiang, v, xi, 175
Hsu, Robert C., 172, 186
Hu, Mengzhou, 168, 179
Hua, Gua-feng, i, 1, 45, 47, 94
Hua, Guozhu, 179
Hung, Yuanpeng, 179

Ishikawa, Shigeru, xi, 10, 42, 57–8, 61–2, 103, 105, 107, 164, 169, 173–5, 186

Johnson, E., 186
Johnson, G.E., 167, 186
Johnson, Virgil A., 173, 186
Jowett, J.A., 186

Keesing, Donald B., 186
Khan, Amir U., 186
Khan, Azizur Rahman, 167, 186
Kojima, R., 187
Kueh, Y.Y., 187

Lalkaka, Dinyar, 165, 187
Lalkaka, R., 187
Lardy, Nicholas, 187
Lee, Eddy, 167, 186
Lele, Uma, 165, 187
Li, Chengrui, 179
Li, Quangshang, 16
Li, Xuezeng, 166, 169, 182
Li, Yu, 123, 175–6, 179
Liang, Chuanyuan, 165, 179
Liang, Liguang, 165, 167, 179
Liang, Wensen, 29, 165–6, 179
Liao, Jianxiang, 180
Liao, Jili, 170, 180
Lin, Qi-hui, 181
Lin, Zili, 180
Lippit, Victor, 166, 187
Liu, Wenduo, 180
Liu, Zhangshan, 180
Liu, Zhongyi, 180
Lu, Zhangshan, 180
Lui, Wenwei, xi, 180

McFarlane, Bruce, 62, 169, 190
Macrae, John T., 187
Maier, John H., 187
Mao, Zedong, i, 30, 45, 53, 68, 94, 99, 130–1, 169, 180
Maruyama, Nobuo, 187
Meisner, Mitch, 174, 189
Morawetz, David, 187
Morehouse, Ward, 171, 187
Myers, R., 187

Nagao, Masafumi, 170, 184
Nai-Ruenu, Cheu, 187
Nolan, Peter, 187

Onoye, E., 187
Orleans, Leo A., 168, 187

Pairault, T., 188
Peng, Xianchu, 48, 180
Perkins, Dwight R., xi, 172–3, 188

Qi, Zong, 180

Rawski, Thomas G., 106, 169, 171, 174, 188
Ren, Jianxin, 171, 180
Reynolds, Bruce L., 171, 188
Riskin, C., xi, 29, 166–8, 175–6, 188
Robinson, Austin, 167, 176, 188

Saith, Ashwani, 167, 185
Schrain, Stuart, 189
Schran, Peter, 174, 188
Selden, Mark, 166, 187
Sen, Amartya, 60, 167, 188
Sethuraman, S., 176, 189
Seybolt, Peter J., 188
Shen, Xiaoli, 174, 181
Shirk, Susan L., 188
Sigurdson, Jon, xi, 71, 77, 131, 167, 170–1, 176, 188–90
Simon, Denis Fred, 171, 189
Sinha, Radha, 168, 176, 189
Solinger, D., 189
Stavis, Benedict, 168, 174, 189
Stepanek, James B., 189
Stepanek, Joseph, v, xi, 176
Sun, Fangming, 174, 181
Sung, Tao, 166, 181
Surls, Frederic M., 174, 189
Suttmeier, Richard P., 71, 170, 189

Tell, M., 190
Teng, Weizao, 172, 181
Tuan, C., 174, 189

Wang, Bingqian, 166, 181
Wang, Enkui, 57, 169, 181
Wang, George C., 190
Wang, Leo-bao, 181
Wang, Shouchun, 178
Wang, Shoudao, 167, 181
Wang, Wen-lung, 181
Wang, Wenke, 181
Wang, Yaoling, 181
Watanabe, S., 176, 189
Watson, Andrew, 167, 190
Weiss, Udo, 165, 190
Weisskoff, T.E., 190
Weng, Qiquan, 179
Wheelwright, E.L., 62, 169, 190
Wiens, Thomas B., 29, 106, 165–7, 173–4, 190
Wong, Christine Pui Wah, 190
Wong, John, 190
Wong, Yongxi, 164, 181

Wu, I-chuan, 164, 190
Wu, Jiuxin, 168, 181

Xia, Deji, 173, 181
Xia, Yibai, 181
Xia, Zhen, 166, 181
Xia, Zhen-kun, 181
Xu, Dixin, 165, 168, 172, 181–2
Xu, Kuan, 174–5, 181
Xue, Muqiao, 166, 175, 181–2

Yamada, Saburo, 186
Yang, Chengxun, 182
Yang, Jianbai, 166, 169, 182
Yang, Ligong, 97, 173
Yao, Yilin, 182
You, Yuwen, 182
Yu, Quili, 94

Zeng, Dechao, xi, 182
Zeng, Qixian, 168, 182
Zhai, Yao, 172–3, 182
Zhan, Wu, 182
Zhang, Baomin, 182
Zhang, Shugyuang, 166, 182
Zhang, Yan, 190
Zhang, Yulin, 182
Zhang, Zhuoyuan, 179
Zhao, Lükuan, 168, 178
Zhao, Ziyang, 38, 89, 178
Zhou, Jizhong, 170, 182
Zhou, Yong-Quan, 181
Zhu, Lunkun, 182
Zijing, Shang, 93

Subject Index

Academia Sinica, 67
Agricultural Bank of China, 14
agricultural machinery, 16, 63, 95–9, 101–3, 111–12, 114, 122, 129, 141, 161–3, 172–3, 178–9, 181–2, 186
Agricultural Machinery Research Institutes, 80, 82, 161–3, 172
agricultural pricing, 31–4, 106
agriculture
  gross value of output, 9–11
  growth of total output, 8–9
  investment in, 5, 61
  low productivity in, 10
  per capita output in, 10–11, 101–2
  price changes in, 28
  research in, 68
  responsibility system, 12–13, 34, 45–9, 55, 68, 95, 99, 108, 167, 168, 180, 182, 185, 186
agriculture–industry–commerce combines, 40, 53
aid
  agreements, 89
  Soviet, 60–1, 83
  supplier of, 83, 89
American Rural Small-Scale Industry Delegation, 94–5, 131–2, 172–3, 188
American Wheat Studies Delegation, 106
Anhui, 105

Bank of China, 86
Baoshan iron and steel plant, 92, 172
Beijing, 14, 52, 76, 147, 151, 161–2, 165–6, 168
brigades, 34, 36, 40, 42, 50, 67, 77, 89, 111, 113–15, 117, 120–1, 124–6, 128, 133, 146, 167, 174–6, 180–1

capital
  accumulation, 41, 65, 92, 156
  foreign, 93, 136
  goods, 70, 83, 91, 170, 184
  intensity, 58, 61, 63, 113
  investment, 92
  labour ratio, 59–63
  output ratio, 59–62
Central Institute of Scientific and Technical Information, 75
centralisation
  of science and technology, 80–1
  policy of, 34, 163
Changsha, x, 15–16, 25–7, 54, 63, 74, 76, 85–6, 121, 127, 147–52, 160–1, 164–6, 168, 170–2, 175, 177, 179–81, 184
Changsha International Seminar, x, 78, 171, 177, 180–2, 184, 187
Chinese Academy of Sciences, 68, 72, 78–80
collectives, 3, 5, 52, 53, 108, 113, 148, 179
collective farms, *see* collectives
commerce
  goods procurement, 24, 28
  department, 27, 34, 39
  sector, 29–30
  system of ownership in, 25–6
commodities
  pricing of, 27–9
  terms of trade, 28–9, 188
Commune and Brigade Enterprises, Bureau of, 76, 111–14, 127–9, 136
communes
  organisational changes in, 39–40
  group motivation in, 41
  problems of, 40
  productive functions of, 40–1
  rural industry, 99, 122
compensation trade, 79, 86, 153
congress
  National People's, 38–9, 52
  Hunan People's, 16, 24
construction:
  average labour productivity in, 23–24
  companies, 22
  employment, 22–3
  share of investment in, 23, 145

projects, 83, 86, 145, 160
contracts
compensatory, 86–7
foreign, 86, 91–92
specialised work, 57
for technology imports, 91–92
cooperative
credit, 14, 165
marketing, 26, 41, 49
production, 84
shops, 25–6
crops
agricultural, 2, 8–10, 12, 106–7, 173–4
cash, 32, 53, 106–7
industrial, 12–13, 99, 173
multiple, 108
Cultural Revolution, 3, 17, 25, 40, 55, 58, 62–3, 75, 79, 83, 130, 170, 190

decentralisation
of decision-making, 30, 78
of economic management, 39–43, 135
of foreign trade, 85
of production, 47
of R and D, 66
demonstration effect, 131
distribution, 25, 34, 41

Economic Reconstruction and Planning, Committee of, 129–30
economic transition, 30, 133
education and training, 52–3, 68, 74, 76–7, 96, 99, 105, 122, 154–57, 170, 175, 185
employment
and mechanisation, 105–10
by application and examination, 54
collective, 5, 51
creation, 5, 50, 168, 170, 182
new policy on, 52, 54–5
of urban youth in rural areas, 53
sectoral distribution of, 3, 6
self, 51–2
state, 3, 51
system, 51–2, 55, 168, 180
urban, 3
youth, 3, 52–3, 55
enterprises
administrative organisation of, 127–8
commune and brigade, 14, 39, 112, 114, 116–21, 124–7, 130–2, 142–3, 168, 174, 180, 182
competition between, 78, 123, 133
creation of, 129–30
growth of, 117–18
financing of, 126–7
large-scale, 123, 134
rural, 42, 114
some economic facts on, 114–15
small-scale, 42, 62, 118, 121, 123
exports *v.* employment, 136–7

factor endowment, 131, 137
factories, 54, 65, 67–70, 87, 122, 124, 128, 134, 135, 137, 148–58, 174
farm machinery, *see* agricultural machinery
Farmers' Bank, 126
First Five-Year Plan of China, 29, 59, 83
Foreign Trade, Hunan Bureau of, 85, 129
forestry, 2, 11–13, 15–16, 96, 113
Four Modernisations, 21, 56, 63, 91, 165, 167–68, 171, 173–4, 177, 183–4, 187–90
fuel pricing policy, 98

Gang of Four, 56, 165, 167, 175, 184, 190
government, 42, 45, 74, 78, 85, 96, 105, 122–6, 132, 134, 136, 150, 152, 159, 176
grain output, 5, 8, 10, 101, 140
Great Leap Forward, 35, 58, 60–3, 66, 75, 83, 122, 130, 134
group motivation, 41–2
Guangzhou, 87–8, 90

handicrafts, 14–15, 36, 48, 111–12, 137, 188
Heilungjiang, 10, 68, 124
Hunan
agriculture, 5–14
area, 1, 5
climate, 2
employment pattern, 54–5
industrial structure, 14–17
population, 1–3
Hunan Agriculture Department, 9
Hunan Commerce Bureau, 24, 27
Hunan Labour Bureau, 3, 52, 54, 157
Hunan Planning Commission, 130
Hunan Trade Corporation, 84–5, 146, 151

import-substitution, 91, 93
incomes
collective, 149

household, 13
rural, 13, 43
India, 42, 49, 101, 103, 136, 176, 183
industrial growth
investment in, 21–2
of Hunan, 17, 19
of Central-South Provinces, 17, 19
rate of, 19, 22
industrial output
growth of, 17, 19
of light industry, 20–1
of heavy industry, 20–1
share of, 17
value of, 18, 144
Industry and Transportation, Hunan Conference of, 16
industries
competition between rural and urban, 123, 134–6
defence, 36
five small, 42, 134
light *v.* heavy, 21, 37–9
new measures for promotion of light, 35–7
rural, 188
small-scale, 42, 187–9
by levels of administration, 113–15
information, dissemination of, 75–9, 183
Inner Mongolia, 124
innovation, 61, 66, 68–70, 75–7, 89, 91, 154, 169, 171, 184, 188
investment
allocation and technology choice, 7, 56–8, 172
coefficient of economic effectiveness of, 58
employment ratio, 63–4
funds, 42
import content of, 83, 92
inducement mechanism of, 61
industrial, 21–2
in scientific research, 71–2
per worker, 34
rate of, 35, 65, 92
reallocation, 34–5
share of foreign, 87–90

Japan, 84, 151–2, 163, 173, 186
Jiangsu, 10, 40, 105, 179
Jilin, 10

Korea, Republic of, 101, 103–4

labour
costs, 108–9, 174
displacement, 94–5, 154
division of, 35
migration of, 51–5, 122
policies, 188
remuneration, 43–4, 188
service companies, 53, 55
supply and training, 121–2
supply price of, 41
land
arable, 5, 8, 95
irrigated, 5, 8, 9
liaoning, 14
licensing
agreements, 86, 152–3
fees, 98
loans, 14, 126
local technology systems, 68, 77–8

Machine Industry, Bureau of, 76, 80, 130
machinery imports, 63, 84, 92
manpower utilisation, 32
marketing
centres, 41
companies, 128
rural, 187, 190
Mass Scientific Movement, 80
mechanisation
and employment, 105–10
and rural industry, 99–100, 182
by rural institutions, 95–6
extent of, 98–105
implications of, 96–8
policy of, 94–6
multiple cropping
index, 106–8, 110
intensity, 108
and labour requirements, 106–7, 110
systems, 104

National Agricultural Conference, 94
Nanjing Petrochemical Plant, 92
National Conference on Agricultural Mechanisation, 94
National Ministry of Machine Building, 84
Ningxia, 68, 123

OECD, 170, 187, 189
ownership
collective, 25–6, 45

levels of, 42
state, 25
private, 25–6
system of, 25–6

Pakistan, 157
People's Construction Bank, 87
peasants, 77, 162
Philippines, 101, 103
planning, 31, 35, 38–9, 42, 56–7, 65, 67, 92, 121, 128–9, 179, 185, 187–9
plant conversions, 38
policies
coordination and specialisation, 35
distribution of surplus raw materials, 35
joint operations, 35
linking coastal and inland regions, 36
new economic, 17, 38, 42, 133, 162, 181, 183
new industrial, 37
pricing, 32–3
uniting the military and civil sectors, 36
population
and employment, 2–5
of Hunan, 2, 4
of Central–South Region, 3–4
porcelain, 16–17, 36–7, 69, 121, 123, 146–47, 154
post-Mao period, 29–30, 32, 56, 58, 63–5, 75, 79, 83, 91, 94, 99, 108, 130–1, 135, 187, 189
price
incentives, 27, 32
reductions, 27
three-tier system of, 27–8, 34
private
farms, 40
households, 40, 45, 47
plots, 43, 45, 47
sector, 42
procurement price, 28–9, 32–3, 106, 125, 159
production
costs, 42–3, 108–9
small-scale, 113, 135, 168
teams, 27, 29–30, 39–40, 42–4, 49, 115, 165, 167–8, 175
purchasing power, 15, 34, 131

raw materials
competition for, 121, 123–4, 134–6
supplies of, 122, 125
research and development
centralisation of, 66
expenditure, 68, 71–72, 155–6, 158, 170, 189
linking with production, 67–8
two-legs strategy in, 66–7
within industry, 69–71
responsibility system
adoption by provinces, 46
and growth of incomes and productivity, 47
commercial, 51
diffusion of, 44–5
industrial, 50, 55, 168, 185
problems of, 47–8
types of, 43–5, 168
rural
communes, 27
diversification, 11–12
incomes, 13
trade and marketing, 26–7
fairs, 26–7, 55
to urban migration, 55
rural industrialisation
case studies of, 146–50
concepts and definitions, 111–14
employment and output growth in, 115–18
financing of, 126–7
future of, 130–2
input, finance and organisation of, 121–8
and farm mechanisation, 99–100
rusticated youth, 3, 53, 148–9, 188

savings
household, 126
deposits, 14
rural, 13–14, 41
rate, 65
science and technology research groups, 80
Science Fund, 72
scientific manpower, 66, 72–5, 154
scissors' gap, 29
service trades, 24–5
Shanghai, 14, 35, 37, 63, 70, 151, 154, 158, 166, 182
Shanghai Handicrafts Industry, Bureau of, 35–6
Shaoyang, 9
shortages

of consumer goods, 34, 37
of foodgrains, 121
of raw materials, 35, 134
of scientists and technicians, 74, 154
Sichuan, 40, 45, 47
sideline activities, 12–13, 28, 42–3, 48–9, 150, 166
state
council, 75, 84–5, 132, 175–6, 178, 180
regulation, 124, 127, 176
sector, 25, 113, 115, 129
State Economic Commission, 79, 87–8
State Planning Commission, 51, 79, 88, 125
State Science and Technology Commission, 76, 79, 81
subcontracting, 114, 124–5, 136
Suzhou Prefecture, 108

taxes and profits, 127–8
technological
dualism, 60, 130
transformation, 56, 91
technology
administrative organisation of, 79–82
capacity to assimilate, 66, 136, 171, 189
choice and investment allocation, 56–66
diffusion, 75–9
imports and foreign investment, 83–93, 171
indigenous, 57, 89, 91
labour-intensive, 59, 60, 65, 177
transfer, 68, 78–9, 83, 170–1, 183–4, 189
unpackaging of, 91
variations over time, 58–9
Textile Industry, Hunan Bureau of the, 159
Thailand, 101, 103–4, 157, 162
Tianjin, 14
training, *see* education and training
transport, 5–7, 98, 114, 116, 117, 158

UNCTAD, 70
UNIDO, 87, 90, 187–8
United Nations, 171, 190

vegetables, 12, 33
villagers' committees, 40

wage
goods, 41
system, 40–1, 54
Walking on Two Legs, 60, 62, 66 130, 133, 186–7
work points, 50, 54, 146, 174
World Bank: 96, 173, 190
Wuhan, 164

Xiantan, 54, 84–6, 148, 155, 157, 160
Xinjiang, 124
Zhejiang, 105
Zhu Zhou, 12, 28, 54, 121, 124, 146, 153, 155